# ON THE ROCKS

# ON THE ROCKS

Text and photographs by Bryan Nelson

artwork by John Busby

LANGFORD PRESS 2013

Langford Press, 10 New Road,
Langtoft, Peterborough PE6 9LE

www.langford-press.co.uk

Email sales@langford-press.co.uk

ISBN 978-1-904078-56-2

Design origination and typeset by
MRM Graphics Ltd, Winslow, Bucks
Printed in Spain

*Previous page: Magnificent frigatebirds displaying for a mate*

# Contents

*Aldabra kestrel*

*Dimorphic egret*

For dear June, whose life our adventures enriched but whose wider enthusiasms, such as other people and esoteric branches of self-awareness, my narrow interests shackled.

*Bryan on the Bass*

# ACKNOWLEDGEMENTS

Trips like the ones that make up this book are oiled by lots of people, many of whom are mentioned where they crop us (apologies to those I have missed) but a few have been pivotal. Those whose claims crowd in on me are acknowledged below in no particular order.

I was hugely helped by Sir Hew Dalrymple who sort of loaned me the Bass and its gannets for many years and encouraged me no end. The inimitable Marr family of North Berwick who almost adopted me. Dave Powell who lent me his Christmas Island and with it the charismatic Abbott's booby. Ron Fisher and Chris Robertson of New Zealand. My ancient (!) friend and artist John Busby for endless help, inspiration and his wonderful illustrations for all my books. My friend, artist and photographer Bas Teunis. Callan Duck for the aerial photo of the Bass. And, of critical importance, Aberdeen and Oxford University Zoology Departments, and the Phosphate Commissioners of Christmas Island.

*Argo*

*Swallow-tailed gulls*

Prologue

# A TASTER

Alarm and excitement, but mainly alarm. We had been living on our sun-baked strip of lava, just the two of us and the seabirds, without sight of humans for months on end and a rather strange paranoia seemed to be taking hold. A white speck had appeared on the rim of Darwin Bay, on Tower Island, in the Galapagos. As it crept in it seemed uncertain, hesitant, but eventually it dropped anchor and soon attracted hordes of red-footed boobies which would have readily festooned its rigging but there was precious little of it and a booby's webs are not well suited to slim wire stays. A mere 21' long, 'Popeyduck', Cornish for puffin, was a toy boat in a rather big bath.

*Red-footed boobies on the* Beagle

"Better get dressed, I suppose". Living on our own, clothes had become unnecessary in this benign climate except as protection from a burning sun or thorny scrub. It was both practical and pleasant to go without for we were often in the lagoon or swimming out to our small wooden rowing boat that saved us a hard slog over razor-edged lava. Our life revolved around seabirds, especially the boobies and frigate-birds nesting in hundreds behind our faded green tent on a tiny coral beach. We did the rounds every day without fail, checking nests as routinely as any office drudge checking invoices, for Tower was our workplace and the birds every bit as demanding as a tyrannical executive though a lot more charming.

*The camp on Tower Island*

Our unexpected visitor eventually plucked up courage to row ashore; a thin, almost emaciated figure, stooping and dressed in ragged shorts and a Francis Chichester cap that had seen service under the engine hatch. Bill Proctor, a late-middle-aged Cambridge graduate yachtie had dropped-out. Abandoning a suited-and-bowler-hatted career as a senior Civil Servant in charge of ancient monuments, he built himself a Laurent Giles designed yacht in his garage and set off alone to sail around the world. This was 1964, long before sailor girls like Ellen McCarthy were even thought of and to us it seemed miraculous that a frail and scholarly government official had risen from his office chair, tidied his desk and said 'cheerio' before sailing off to face danger, privation and deep loneliness. Maybe it seemed equally surprising to him, coming to an uninhabited speck of land in the equatorial Pacific only to find a couple of bird watchers squatting on a shingle beach barely above high water mark.

"I nearly turned back when I saw your tent" and you can easily see why he might have been wary. The Galapagos has attracted some weird characters and you never know who might be a highly unwelcome visitor. Naturally he was curious about us and luckily for my self-esteem seabird biology was perhaps the only subject that this erudite man knew little about. I disgorged a few tasty morsels about sibling-murder in the masked booby and infanticide in frigates in exchange for his learned lectures on bee-keeping, ancient monuments, modern poetry and numismaty. Perhaps he found any old ear better than talking to his slop bucket at sea.

His predatory eye, already glinting at the sight of our paraffin stove, roved to our modest shelf of paperbacks. "Would you care to swap a few books?" We would have been delighted, I think, but the offer was swiftly withdrawn after a brief glance at our rubbish. I never saw his but imagine it covered French philosophy, dissolution of the monasteries under Henry VIII and how to remonstrate in Tibetan to a recalcitrant yak, so maybe we didn't miss much. He didn't manage to wheedle any paraffin out of us but he did triumph in a roundabout way by baking his next batch of bread in our oven. Bill Proctor would have come out top in any Arab bazaar.

Fieldwork takes a lot of time and effort and from the moment we rose from our creaky old camp beds we were seldom idle except for the odd soak in the crystal clear lagoon behind our tent which filled and emptied with each tide. Our newly acquired ex-civil servant was keen to come with us on our booby rounds or maybe he just appreciated a bit of comely female company. Anyway, he tagged along and I doubt whether, when he was building 'Popeyduck' he ever imagined himself duelling with an aggressive booby as it clamped a powerful bill on the nearest bit of his anatomy. But it was one less for us to handle.

After four days Bill Proctor hoisted sail and made his lonely exit from Darwin Bay and out into the Pacific. We never saw him again, nor did we expect to, but he did write from New Zealand, enthusiastic about his travels. He died violently somewhere in the southwest Pacific. Almost certainly he hit a reef at night and the massive breakers smashed little 'Popeyduck' to matchwood. Only small fragments were found and no trace of Bill. It had nearly happened to him on a previous occasion in the middle of a black night, but he had been lucky and clawed his way off. Bill Proctor was no Chay Blythe, no renowned adventurer, just a man of courage and ability who lived out his dream. Maybe we were following our own little vision but it was nowhere near as dangerous or demanding as his. What a pity that he never published his story; it would have been a joy to read.

But I'm well ahead of myself.

*Frigatebirds*

*Gannets hanging in the wind*

## Chapter 1

# THE BASS ROCK

*"Hearken thou craggy ocean pyramid!*
*Give answer from thy voice, the seafowls scream,*
*When were thy shoulders mantled in huge streams,*
*When from the sun was thy broad forehead hid?"*

John Keats 1818 'Ailsa Craig'

The Galapagos islands, made of cinders and lava, are young and our green canvas tent on Tower Island was soon thickly coated with fine ash from eruptions elsewhere in the archipelago. The Bass Rock, too, had a fiery birth but the marks are more subtle. Its precipices of basalt rise so abruptly from the seabed because they are hard rock, formed from the plug of an ancient volcano. The sea surges in vain against this compact clinkstone. The Bass cliffs are angular and sharp-edged, even though less than two miles away across the Forth on the Lothian plain the rocks are softer, fretted and worn - "had the Bass originally been composed of such a yielding tuff as that on which the fortress of Tantallon is erected we would now in vain seek its place amid the waters".

    For several years we clambered around the Bass ringing gannets, sometimes partway down the formidable east cliffs. Now, nearly fifty years on, I wince at our foolishness. It was the same on Christmas Island, climbing trees to reach the nests of Abbott's boobies. June came within a cat's whisker of tragedy when a wooden rung I had nailed to the tree as a ladder, split and gave way. The tree stood amongst needle sharp limestone pinnacles but, amazingly, she fell against a nearby sapling and slid safely down to land on her feet in the only patch of bare earth amongst that array of lethal spikes. I don't know what were the chances of that but it saved her from horrible injury. Again, it was my carelessness. Even after that I regularly climbed high in dead trees

*Tree mallow*

to weigh Andrews frigatebirds. Looking back, it seems crazy but it was the only way to get the information. I fear that my own rough-and-ready approach played a part but it may also be that the 'Oxford experience' contributed something, for in the 1960s there was a 'do-it-yourself' cost-cutting, string and sealing-wax spirit abroad in our Animal Behaviour group under Mike Cullen and Niko Tinbergen. Many of us were hardy souls and Niko himself set the tone.

Well over half-a-century ago I stayed on Ailsa Craig ringing gannet chicks. One

*View to Tantallon*

day I climbed a little way up the rock to ring a few kittiwakes. The nearby fulmars spat their foul oil, which stinks like rotten turnips, and the kittiwakes split the air with their passionate "kitti-wark, kitti-wark". At one point, I stuck. At such moments it is easy for a novice like me to succumb to a desperate reflex, an escape act, rather than just wait for a few seconds to calm down. I had exactly the same feeling one March evening, halfway down the East cliffs of the Bass. I had climbed down alone to check for early gannet eggs. Nobody knew I was there. Indeed there was nobody else on the Rock apart from the lighthouse keepers and they were down in their cosy living quarters. It was a miserably grey dusk, with a bitter wind, and beginning to feel eerie above that menacingly dark sea. I wanted to be back at the fireside in the lighthouse but I entered the wrong crack and it petered out. Suddenly it seemed urgent to get back up quickly, although in fact there was still plenty of light. Then came the urge to take a chance rather than go back down and start again, which was the sensible thing to do. This temptation to take a risk can be a subtle adversary, biding its time and striking at off-guard moments. Young males are notorious risk-takers – responsible for many a tragedy on the roads, at sea and in the hills. The urge may well lie deep in our past and must have brought rewards often enough to have given this tendency a genetic basis. With the ever-increasing security of modern life extreme sports seem to be more and more popular among young people, the risk itself being apart of the attraction.

Kittiwakes:
1. Greeting
2. Well grown young
3. Begging

These experiences are all part of the magic of seabird colonies and it was Ailsa Craig that hooked me. In July 1953 when I first set foot on this mighty Craig, Jimmy Girvan, like his father before him, still quarried the beautiful grey-green granite for the famous curling stones. His wife ran a teashop in a large, rickety barn at the top of the landing jetty. Their goats browsed happily on the short, sweet sward and provided delicious milk for the teashop, now long gone. On that first unforgettable evening I climbed to the top of Ailsa on the precipitous bracken clad slopes among wheeling, cacophanous herring gulls and gannets, black-tipped white crosses against a blue sky. It was Ailsa that shaped the rest of my life. I would like to say it was the Bass, but Ailsa came first. I think it was Hector Boece (1527) who was the first to mention Ailsa's gannets though, like Bass gannets, they must have been known long before.

*Gannet bowing*

Kennedy's Nags, chocolate, sheep ticks and pemmican. On a blazing July day, a friend and I climbed from sea-level to above 1,000m through tick-infested bracken before gingerly descending to the sun-baked ledges and the rotten slabs of rock, dangerously fractured by the explosives which had been used to split the granite. Here, in the heat and the ammoniacal stench from fish and guano we tenderly placed our expensive metal rings, paid for from our own pockets, on the chicks' sturdy black legs at the cost of a criss-cross of lacerations on the backs of our hands, for although their bills lack the steel of the adults' the tips are sharp and wielded with wild enthusiasm. Both then and many a time since I have been surprised to find that despite a coating of guano and black ooze the cuts healed quickly, perhaps disinfected by the ammonia and sulphur. Our ankles and midriffs sported great belts of tick bites, the clegs jabbed with red-hot needles and itchy gannet fluff stuck to our sweaty faces but to us it all seemed magical - what life was all about. Over lunch, a bar of chocolate, we counted our remaining rings and checked our notebooks. Our sustenance for the week, my friend John Leedal's department alas, for he was a hardy lad and bent on making me so, nestled in its tins, pemmican a hard brown wax which hot water magically transformed into a nourishing soup to go with our sliced white pan loaves. These soon sprouted an impressive crop of mucor and penicillin. Since the days of Franklin, the arctic explorer, pemmican has been a famous food. Ours was left over from an expedition to Greenland by John's twin brother, Phil, a geologist.

At that time I had not the slightest notion that within a few years I would be living on Ailsa's famous east coast rival, the Bass Rock. But in 1960 I began what is now half a century's involvement with the Bass gannets. The following year, on a bright day in February, Fred Marr loaded up the ancient "Norah", a sturdy double-ended Fife fishing boat little more than 20 feet long. There were 50 large concrete cones to be cemented onto the Rock in our main areas of observation, to act as markers in mapping the layout of the nests to discover whether gannets stay faithful to their site. In total these delicate little cones weighed 2,238 lbs. And every last one had to be landed on the Rock and lugged up in a rucksack. A few remain in place to this day and I still cannot think of a better way to mark solid rock littered with seaweed and other nesting debris. It wasn't rocket science but it worked. We tried hammering iron pegs into cracks in the rock but the heavy, clumsy gannets soon knocked them over.

When, courtesy of Sir Hew whose family has owned it since 1706, I first set foot on the Bass in July 1960 I lodged in the lighthouse, a tribute both to the keepers, who took a risk with somebody who could have been a walking disaster and the Commissioners who allowed it. I slept in a makeshift bunk in the tiny radio room; the beginning of a great adventure. The lighthouse living quarters were lined with thickly varnished pine and the old-fashioned, coal-fired range and oven fairly gleamed. A large polished copper urn sat to one side of the fire and delivered scalding hot water the colour of concentrated urine 'a dilute tincture of guano'. The water came from a well filled by seepage water which owed its colour to the humus through which it had percolated. A century and a half ago that same well was "full to overflowing with a turbid brown fluid — which had proved the grave of a hapless sheep during one of the snowstorms of winter". Prisoners on the Bass had to drink this corrupted water, sprinkled with a little oatmeal though I doubt that this helped much. In my day this well had a sturdy wooden lid on it! The keepers' outside lavatory at the back of the accommodation block – far enough at midnight in a gale - was flushed with the same yellow water! Baths were taken in a large, shallow, metal pan placed comfortably in front of the kitchen fire. Commodious pine cupboards or 'presses' lined one wall and an old wooden clock with Roman numerals ticked portentously on another. A pressurised tilley lamp hanging from a hook in the ceiling hissed companionably as it shed a mellow light on the scrubbed wooden table. During the long night watches hundreds of letters home were penned. What could the keepers find to say after years on the same old Rock: "Bertie burned the porridge yesterday"; "it's been very cold lately" or "we've been painting the railings". Their sleeping quarters opened directly into the

living room, each narrow cell with two-tiered bunks in solid pine followed the curve of the wall. In the evenings an old black cat with a broken tail slept by the fire whilst the keepers read, chatted, smoked and drank tea. Alcohol was forbidden and this rule was strictly followed; lives depended on it.

Alas, the splendid coal-fired range and oven now lies on the sea-

*Sir Hew Hamilton Dalrymple on the Bass Headland, now covered in nesting gannets*

*Bass Garrison*

bed just off the East cave, dumped there during modernisation although it was still perfect. By the late seventies heating and light was electric, the outside loo had been replaced by a modern bathroom, there was a posh kitchen, colour TV and separate, single bedrooms from which you emerged as straight as when you entered. All that lovely old pine cladding had disappeared, bright synthetic materials bloomed everywhere, the cat and the pinewood clock had gone. Nowadays there are no keepers either. The light is automatic and will soon be abandoned altogether. A hundred-year blip in the old Rock's history will end and the Bass will enter yet another phase.

Back in 1960 I had to get to know the Bass, select gannets for close study and put coloured rings on them so that we could follow their individual lives in the colony.

*Inside our hut, 1962*

Then, early in 1961 came the seminal step of creating our own home on the rock. It was just an ordinary garden shed made of cedar wood, 12 feet by eight, but what a site. We placed our hut within the ruins of the 14th Century chapel which, venerable though it was, sat on the site of St. Baldred's 7th Century cell.

It was equally vital to choose a good position for our observation hide. I wanted to place it near the edge of an expanding group so that I could

Above: *Observation colony with hide, 1960s*

Left: *Colony 1975*

watch the entire process of site-establishment and pair-formation so I put it on the north-west slope with site-establishing birds right under the window, absolutely ideal. It was this group among others that held my cement cones. Today, the whole of that north-west slope is a vast sea of gannets. The site where once my hide stood is now bang in the middle but in the early sixties we could crawl across the hillside and into the hide without disturbing them, though the gulls created bedlam. To provide a base for the hide on that sloping face we built a platform of boulders and turf but even with strengthening stakes it was difficult to stop the whole thing from sliding down the hill, taking us with it.

When there were visitors on the Rock – far too often for my liking – one of us sat in the hide to deflect those who would have entered the colony and disturbed the gannets. If that had happened much laborious work would have been undone and our vital esti-

mate of breeding success unreliable. We tried to answer visitors' questions but some did leave us bemused. The leader of a party from a Roman Catholic seminary asked whether the herring gulls stole gannet eggs to eat or to hatch out. Maybe he had a vague idea that some birds rear another bird's young, like meadow pipits might rear a cuckoo. But the pipits don't steal the cuckoo's egg!

Like Fraser Darling on North Rona and Ronald Lockley on Skokholm, we made the Rock a satisfying home for those few essential years. I have been on Skokholm when the spring bracken was burgeoning and the bluebells bloomed, and on Rona's grassy slopes, and despite Rona's remoteness I know that the Bass is far harder to live on. It is, after all, only a fraction the size of the other two and much of its seven acres is bare rock. In fact the only spot suitable for us to live on was a few square yards of south-facing ground about 150 feet above the sea where the ancient chapel faces Tantallon Castle across nearly two miles of sea. As I said, we put our hut within the sheltering, storm-fretted walls of the chapel and I reckon we crossed that worn old threshold more than 30,000 times. On a fine, sunny morning like those often experienced by the ordinary visitor, who can imagine it on a tempestuous winter's night ----" what time the midnight moon looks out through rock and spray and the shadow of the old chapel falls deep and black athwart the sward". Many a time Fraser Darling's description of Rona ('Island Years' 1944) fitted the Bass exactly "the baffled wind gathers ---- and comes down with a terrific buffet. The hut shakes visibly and tired though we are, and falling asleep between whiles, those dunts of tortured air wake us suddenly every time". We nailed the waterproof felt closely to the hut roof but the wind still managed to get underneath and ripple it.

Our little hut from Halls Ltd may have been modest but it had to be transported to the Bass, lugged up onto the landing stage from the old 'Nora' and then carried 150 feet up the Rock. 'Nora' was not beamy enough to accommodate the sections inboard so they had to lie athwart, overlapping on each side. She came into the inner landing, with its hand-cranked lifting crane, but there was too much swell so she motored round to the outer. Alas, the rolling motion caused the underlying hut sections to slide out from beneath the upper ones. Fred's father Alf, by then over seventy, skipped nimbly from

side to side in a vain attempt to halt the oscillations but with a final de-
risive tilt one section slid gracefully overboard, and a protruding nail
ripped the seat of his sturdy breeks. The rest of the hut followed and
then our humble furniture, bobbing slowly out to sea on an ebb tide. It
was not an auspicious beginning and Alf never took to me after that.

After the excitement I wandered down to the landing in the fading
light. A black frieze of shags lined up on the rocks, a seal eyed me from
a distance, a dark ball on the grey sea, herring gulls wailed and a pair of
purple sandpipers pottered around at the water's edge. This was to be
our back garden. In a few crowded hours the summer visitor, surrounded by others, cannot begin to capture this heady feeling of belonging, of 'ownership' and island euphoria.

We could easily have had to wait weeks for a day calm enough to
lug the hut sections up from the landing to the chapel; the wind caught
them like sails and the steep narrow steps were grimly unsympathetic.
But the following day was flat calm. As soon as I had cleared centuries
of nettles and tree mallow from inside the chapel the Principal keeper,
George Robertson (and no proud prelate rules his roost more autocrati-
cally than a Principal keeper) laid down the floor of our hut on 41 stone
pillars. It was his idea to raise the floor in this way and it saved us from
flooding during more than one torrential downpour when water
coursed down the hillside and through the chapel. And it was George
who supervised the work and passed stout wire hawsers over the roof,
securing the ends to iron pegs driven deep into the base of the chapel
walls. Nor was this precaution excessive, although left to my own slip-
shod ways I might have ignored it. Even within the chapel the west
wind, which funnelled straight up a deep, steep-sided gulley, lifted the
hut roof until the hawsers groaned and the crocks fell off the shelves.
Without George's cables we would surely have lost the roof and then,
inevitably, the walls. The cost of all this labour, meticulously itemised
in copperplate on the back of an old envelope, was, in the EU's debased
currency, fewer than 6 euros.

*The beamy old
'Nora' took our
hut sections
across to the Bass
in 1961*

*Fred Marr, skip-
per of the 'Nora'
and later of
'Sula', who for
nearly 70 years
took visitors to the
Bass. Known far
and wide as 'King
of the Bass'*

13

I had to exercise a degree of diplomacy over our dear little hut. The zoology department of Oxford University had paid for it and Niko Tinbergen had hoped that it would become a permanent base. Sir Hew, on the other hand, had made it perfectly plain to me that it was to be strictly temporary accommodation, specifically for us. Furthermore, we were acutely aware that we must not, emphatically not, have a constant stream of visitors for this made big demands on Fred who had to bring them across, and on the keepers who always went down to the landing to handle the ropes. To complicate matters further, I knew that George Waterston would have liked the hut for the S.O.C. but there was no chance of that. This sort of thing can cause problems; but in this case it didn't, for we removed it and all traces of our occupancy before we left for the Galapagos. How well I remember going alone into the old chapel, empty once again, after everything had gone, and saying "thank you". Silly really.

How can I distill the essence of our life on the Bass? I think there were two pivots: one was an increasingly intimate appreciation of the magnificent gannet and its social behaviour which, previously, had been surprisingly ignored. James Fisher and Ronald Lockley had been famously associated with gannets but almost entirely in connection with their numbers and distribution. Social behaviour remained more or less a closed book despite Julian Huxley's pioneering film and notes (which he kindly gave to me). Gannet ecology, too, was hardly scratched. I wanted to do a lot of scratching. Did anybody know how long it took a gannet chick to grow to fledging age and how much fat it put on? No. How old was it before it began to breed? What was its hatching and fledging success. Could anybody interpret, with evidence, its complex rituals? Again, no. The second essential was the cosy insularity of our self contained and highly focussed life. That it was also delightfully harmonious was a terrific bonus and a huge tribute to June. Inside our little hut the cedar wood's aromatic oils volatilised and this fragrance never faded. Whenever I smell cedar wood it brings back the Bass hut with its cheerful red and white check curtains and all its cosy clutter. We weren't the only ones to find it cosy. Earwigs bred by the million amongst the gannets nests and invaded us en masse. We once found 141 in one small chip-pan. The total cost of our home, workshop, living quarters and library, including transport, building materials, furniture and labour was less than £200, which some people would pay for a bottle of perfume. Homo non sapiens.

These days a field scientist looks to his University, Research Institute or equivalent for everything from equipment down to purely personal gear. And he or she claims away-from-base expenses at rates which may far exceed the money actually spent, just as many MEPs claim expenses much greater than the sum expended. Quite blatantly. Throughout many years of fieldwork I did not claim a single day's subsistence. A bit of restraint could save Universities and the Scientific Civil Service a fortune. But we were more than grateful to accept the help of Her Majesty's Commissioners of Northern Lights in getting our food brought out with the lighthouse keepers' reliefs. These were red-letter days. The keepers' orders were packed by an old fashioned grocer, Coopers of Leith. Somebody in Coopers was a saint in a brown coat. Every item, listed and priced in copperplate handwriting, was painstakingly wrapped in newspaper. Year in year out there came to the Bass as to all the other stations in the Firth of Forth, those sturdy green boxes with stout rope handles and the name of the station – Bass Rock, Bell Rock, Fidra, Isle of May – in bold white paint. I imagined this anonymous perfectionist, a stooping figure with pince-nez, working unhurriedly in an old-fashioned grocery smelling of coffee and cheese, endlessly wrapping and listing. Stores like that are now as dead as the dodo, killed off by the all-conquering supermarkets.

Lighthouse reliefs came once a fortnight; so our paraffin fridge was a great boon although generations of keepers had managed with a limed, fine-mesh larder which still hung in the bottom of the lighthouse tower. Our safari camp beds came from Milletts, the folding wooden chairs were Woolworth's best at 99p each and our sturdy deal table cost 60p secondhand. Hot water came from an old copper urn with a brass tap (a wedding present) which sat on top of a flat wick paraffin oven. That simple little oven baked many a batch of Yorkshire puddings and eventually came with us to our desert island in the Galapagos. We interred its bones in a deep crevice on Hood Island, well out of sight. By now it will have rusted away. We drew our washing water from the old Garrison well in the garden and carried it down to the hut in a plastic dustbin, a most unyielding chore. We bathed in the sea, (in truth we often merely dipped our backsides) or else in a plastic bowl and disposed of our chemical toilet in holes laboriously picked out of the stony slope above the chapel. Even more than Her Majesty, we ran a tight little ship. Tight it may have been but, perhaps surprisingly, we rarely tried fishing. Fred often threw a few mackerel and the odd crab our way but all we ever caught for ourselves were saithe which we found tasteless.

Every relief day was a birthday: fresh meat, fruit and vegetables, new bread, cream and chocolate digestives. Often enough I came down from the hill after a session in the draughty old hide to find June, flushed and thick, glossy hair awry, in the midst of pies and scones which were cooling on the floor. Potatoes and vegetables simmered on the stove, chops sizzled and the paraffin oven was going full blast. Aromas blended deliciously with the scent of cedar wood. At the tail end of the fortnight  herring gulls' eggs and chips made a good, cheap meal. We preserved the eggs in waterglass and they made excellent cakes. Herring gulls are persistent layers and continue to replace pilfered eggs until their pigment glands become exhausted, when they produce pale blue eggs rather than the usual green or brown heavily blotched with black. Rabbit pie was another favourite, though there were no rabbits on the Bass during our time. Until 1958 they had been abundant, replacing the sheep which used to graze on the fine turf. Bass mutton was famous. Unfortunately, although fescues and bromes still clothed the summit and south-facing slopes the north and west faces were covered with coarse and fleshy Yorkshire fog, which thrived on the disgusting mess made by the herring gulls. The summit turf, deeply dissected, looked suitable for puffins but old gaudy-neb never penetrated beyond the garrison walls where a few pairs nested, as they still do, in the deep cavities between the ancient stones, emerging in twos and threes to sun themselves. Bass puffins give the impression of hanging on by the skin of their beaks in this unlikely breeding place, but elsewhere in the Forth there are tens of thousands. On the nearby Isle of May they increased from a mere handful in the 1950s to an estimated 69,000 pairs in 2003, though they have since decreased. Many of them must be immigrants from other colonies and even the small group on the Bass which one might think

*The puffin, the "pin-up" of the seabird world*

was self sustaining, may receive immigrants. I picked up a dead puffin which had flown into a cable and found that it had been ringed on Grassholm off the Welsh coast of Pembrokeshire – quite a journey by sea whichever way you fly.

In those halcyon days of the early 1960s, and the older I get the more they shine, the ancient Garrison garden was tended devotedly by the keepers. They grew wonderful tatties, nourished by seaweed and guano from old gannet nests and often took their produce ashore, along with the fruits of their labour in the workshop. Perhaps it would be a jewelry box decorated with shells or, in George's case, a superbly crafted writing desk, lustrously french-polished. He wrapped it carefully in sacking and sold it ashore for a few pounds – a real bargain. Television's deadening hand soon put paid to all that nonsense.

As incomers to the Bass we did not merit a patch in the valuable walled garden, but we hacked away at a small area outside and cleared another patch near the summit,

*The walled garden in 1962 when carefully tended by the 'Keepers'*

*Garrison garden in the 90s when overgrown and invaded by gulls*

17

*Redstart*

looking out to the Isle of May, about 7 miles to the north. It was a pretty thankless task. We planted Midlothian earlies but the yield was East Lothian late, and meagre too. Outside the garrison garden dense clusters of small wild daffodils, centuries old, flowered each spring, beneath an incredibly gnarled and ancient elder tree which disappeared in 1999. More than a century ago Miller wrote: "the garden, surrounded by a ruinous wall, when seen in the genial month of June, 1842, bore among the long lank grass — a delicate sprinkling of garden flowers grown wild.". They are long gone, but on a bright, calm spring day (miserably rare) with the singing rock pipits parachuting to earth, early willow warblers flitting along the lichened walls, gulls above and the excited clamour of gannets all around, it was a blissful place.

Those late winter days held a special magic. A new field season was afoot, the Rock was shaking off winter's dead hand; the Bass, when fast frozen and powdered with snow, leaves ample room for improvement.

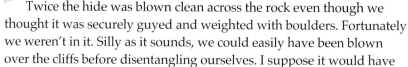

Snow buntings often foraged amongst the old gannet nests and during the spring migration anything could turn up. From mid-March onwards large falls of robins, thrushes, blackbirds, redwing, bramblings and goldcrests followed periods of mist and easterlies. A little later came wheatears, redstarts, willow warblers and ring ousels. Raptors came with the migrants. An unlucky peregrine failed to pull out of a massive stoop and jammed itself head-first into a crack in the rock. It won't do that again! Another doughty falcon, the merlin, in hot pursuit of a house sparrow, followed it from higher on the Rock right down to the battlements in a breath-taking dive before carrying it back up to the chapel to give thanks, pluck and eat. At Azraq in Jordan I once saw one of these great little falcons chase a red-rumped swallow and pluck it out of the air with one taloned foot.

*Wheatear*

Every spring thousands of common gulls passed high overhead, almost out of sight; their querulous calls floating down, barely audible. This happened within a day or two of April 18 on each of three consecutive years. Then would come that accursed wind. We expected gales in Spring and Autumn; they were part and parcel of the seasons, but blizzards of hail and snow in June was a rotten deal. So were gales or near-gales that went on and on for a couple of weeks when the radio insisted that we were enjoying a fine, settled spell. Small islands are much colder and windier than inland areas and we hated those prolonged westerlies. They soughed and roared without respite, buffeting us off balance, mocking any attempt to use binoculars or write in notebooks; eye-watering, nose-dripping irritants.

*Merlin chasing a sparrow*

Twice the hide was blown clean across the rock even though we thought it was securely guyed and weighted with boulders. Fortunately we weren't in it. Silly as it sounds, we could easily have been blown over the cliffs before disentangling ourselves. I suppose it would have been a unique way to die. At night the wind exploded against the hut, twanging the stout wire hawsers that secured it and threatening to lift the roof. Listening to the surge of the sea in the blackness I sometimes thought of the single-handed sailors who go down into the wastes of the southern ocean, into the screaming 'fifties. Imagine being alone in the blackness off Cape Horn in those awful seas that engulfed more than one fully-crewed clipper on the tea-run from Australia to London. There they lie, in their eternal solitude.

Just below the old chapel at the top of cable gulley a deep hole has been blasted out of the rock. More than a century ago it held run-off water for mixing the vast amount of concrete for the path which runs from the lighthouse compound, over the top of the rock and down to the foghorn which faces the Isle of May. Over the decades this pit had filled with rocks and mud. When I first clapped eyes on it the docks and nettles were waist high. I blithely began to clean it out thinking it might make a nice little plunge pool. What a job! .Every lunge of the spade hit a stone. Every bucketful of stinking ooze had to be squelched out through the clinging mud and slung down the gulley. It was hard labour but when the hole had been cleared out down to the bare rock it gradually filled with deep yellow seepage water like that in the old Garrison well higher up the rock. Still, it was a pool and had you chanced unexpectedly on the scene on many a summer's day you might have wondered at the sight of Adam and Eve leaping blithely into this sinister hole. Now it is again solid with debris.

Rather to my surprise my diary of those Bass days insists that we

*The spread of gannets on head-land around the old chapel, which had no gannets anywhere near in the 1960s & 1970s*

*Kittiwakes
fighting, by
Bas Teunis*

loved the simplicity of the life. If we truly did, we have dismally failed to learn the lesson for, since then, our lives have become ever more complicated and our pile of non-essentials much greater. But I know what we meant. There was a simple rhythm to daily life, a small circle of friends, a remoteness from piffling distractions. There was the satisfaction of identifying deeply with an enthralling and romantic island and the opportunity to get under the skin of gannet behaviour in a way that had never been done before. And you couldn't ask for a more spectacular bird than the gannet. I was silly to write, as I did, that I could happily live for ever on the Bass, but at the time I believed it. I suppose I was swept away by romanticism. Fraser Darling was similarly affected by his island-going but he progressed far from the simple life on Rona and the Treshnish Isles about which he wrote so lyrically, to a love of antiques and fine wine. But in his early days he was marvellously able to enthral his readers with the delights of life on a small island, the great outdoors and his own intuitive, naturalistic kind of research. Although he approached fieldwork scientifically he had a strong aesthetic appreciation. The boggy patch near their tent on North Rona had to be left undrained even if it was a bit of a nuisance, because a snipe regularly fed there and to Darling it was unthinkable that it should fly in to feed only to find that its patch had gone. It was the joy of living on Rona quite as much as the scientific work that drew him there. I can easily relate to that, but his approach to animal behaviour, as distinct from ecology, doesn't inspire me. He was not really an ethologist but he was certainly a wide-ranging synthetic thinker rather than a precise analyst. Maybe at times he strayed beyond the strict boundaries of science but for every Fraser Darling there are a hundred conventional scientists.

The Bass has been a central part of my life for nearly fifty years. When we packed our ten wooden boxes for the Galapagos we had by no means deserted the old Rock; we would be back many times. If one can feel gratitude towards a lump of basalt, 'a craggy ocean pyramid', then I do. Over the years it has changed in ways I never could have envisaged when I first set foot on it in 1960. I have seen it in all its moods, lived totally alone on it in tempestuous weather, watched the seasons and the gannets come and go " the auld craig wherein is great store of soland geese". It will be there "when man is redeemed or redundant"; probably the latter.

*Gannets bowing*

*Mass of diving blue-foots*

## Chapter 2

# CASTAWAYS

*"The thrill of one's first desert island is quite indescribable and the fascination of the barren Galapagos inexplicable. A land pitted by countless craters and heaped with a myriad mounds of fragmented rock; gnarled trees bleached by salt-spray, …. grotesque cactus stiffly outlined against black lava – this picture cannot convey the feeling of mystery that led us on and on, wondering always what lay just beyond the next barrier of sombre rock"*

William Beebe: 'Galapagos: World's End' 1924 G.P. Putnam's Sons. New York

From the basalt of the Bass to the lava crust of Quita Suena – Devil's Island, alias Tower or Genovesa, our new home. From gannets to boobies, herring gulls to lava gulls, hut life to tent life. The selfsame paraffin oven and refrigerator which we packed away on the Bass we now unpacked half a world away on Tower Island.

Since Darwin's brief but mould-shattering visit to the Galapagos in 1835 nowhere on earth has been more celebrated than this forbidding group of volcanic rocks and islands straddling the equator in the eastern Pacific, a moonscape of buckled black lava. Well might it be called "world's end". Escape to a desert island is a daydream which for most people would quickly turn into a nightmare. Lucy Irvine's manufactured sojourn on a tiny south Pacific island brought her TV acclaim, a film version and a book devoted mainly to her love affair with the sun and her love/hate relationship with her rather odd husband-of-convenience, but it was basically an escapade. The reality, when there is sustained and demanding field work and no contact whatsoever with the outside world for months on end, is rather different.

The world and the Galapagos were very different in 1963. Nowadays you simply could not do what we did. There is now huge pressure on the Galapagos Islands from tourists – which hardly existed in 1963 – and from the thousands of recently arrived, mainly urban Ecuadorians. The free-and-easy, self-sufficient days are long gone. The sweep of Darwin Bay which met us every day week after week and month after month, thankfully empty of human life, is now visited several times a week by tourist boats. Passengers (supervised) troop ashore on 'our' little beach that in 1964 knew only a few sea lions, the sooty black lava gulls, the lovely swallow-tailed gulls and us, in our tent.

In 1964 the Charles Darwin Research Station on Santa Cruz was officially opened; a momentous occasion hailed with fanfare and great international goodwill. Ecuador invited many distinguished scientists from Europe and North America to visit the islands for the inauguration, to fall under their spell and, it was hoped, fund future research. During this week-long jamboree a number of these eminences wove an erratic course into Darwin Bay on board an Ecuadorian frigate with distinctly idiosyncratic steering gear. It didn't have to come in backwards, as the old Cristobal Carrier once entered Academy Bay, but it was hardly Royal Navy stuff. Even the attendant red-footed boobies were a bit non-plussed. We were eagerly expecting to see NikoTinbergen, the famous Dutch ethologist, who had planned to stay on Tower with us for a week or two, but his distinctive silver head was nowhere to be seen. An accident to his wife Leis made him cancel and there had been no way to let us know. I was surprised to see what a crowd of doddery ancients those academics were but now that I am more than equally challenged I don't find it at all remarkable.

Our own little venture, hardly to be dignified as an 'expedition', was very much a shoe-string affair with no official backing. We organised it ourselves during our final hectic year on the Bass – no simple task in those days of snail mail. My letters to Ecuador enquiring about transport to the islands and such like simply went unanswered; not in the least surprising considering that with typical British hubris I wrote in English. In retrospect it was perhaps fortunate for if we had known how erratic were the old 'Carrier's' sailings, and how unreliable, we might well have given up. In the event we were lucky beyond belief.

To mount a year-long expedition to two widely separated, waterless and uninhabited islands hundreds of miles from the mainland and thousands of miles from home would cost a fortune today. It would require the backing of a University or a Government-funded Institution. We had no such sponsorship and very little private money. We received some hugely appreciated help from the American Frank M Chapman Foundation, set up to help young field workers like us and administered by the wholly admirable Dean Amadon. With absolutely no fuss this fund helped with travel costs and, vitally important, equipped us with a hand-cranked cine camera with which I took invaluable footage of booby and frigate behaviour. Many of the line drawings in my monograph on the Sulidae came from those films.

Mainly, though, we relied for our Galapagos trip on thrift, good planning and frugal living. All our equipment was well worn from hard years on the Bass Rock. The 11 wooden chests in which we carted everything by sea half way across the world were made by the Bass lighthouse keepers. They had strong rope handles and corners neatly protected with metal cut from Tate & Lyle syrup tins – the ones with a dead lion, a cloud of bees and the motto "out of the strong comes forth the sweet". Most organisers of a year-long expedition to a remote island would have spent our budget on stationery and postage alone. Some years later I happened to meet an American, Paul Colinveaux, who had worked in the Galapagos. To my annoyed surprise he was rude and aggressive, largely because he considered our expedition set a bad example by making it harder for folk like him to raise cash, especially if you want to hire a plane to fly things out. Quite so; the American approach. It was a new idea to me and clearly I had much to learn. We paid our way and went home with some wonderful results. But Empire Builders, whether political or academic don't work like that.

But I'm far ahead of myself. I haven't even got us to our desert island yet. First we had to cross the Atlantic to New York en route for Guayaquil in Ecuador. Half-way across we ran into a genuine Force 12 hurricane which forced Cunard's 'Queen Elizabeth' to heave-to and that doesn't often happen. Her great bow dipped ponderously into the deep troughs before slowly rising, until it pointed high above the horizon. When she was down, the seas towered high above the top deck and when she was up, the valleys and slopes below smoked with spray, whipped off horizontally in searing sheets. And yet there were kittiwakes, mere bundles of feathers, riding out this elemental fury. We encountered hundreds of them, flying with the wind; not that they had any choice! Among them were two juveniles with their distinctive black collars and wing bars, perhaps siblings, now far from their noisy colony. And there were fulmars, shearwaters and a single guillemot. About 100 miles east of the N. American coast we encountered several Canadian gannets, indistinguishable from ours. There may be slight interchange between the two populations and it seems surprising that there isn't more, using Icelandic gannetries as stepping stones. Perhaps there is, for it would be impossible to detect simply on plumage or measurements.

In New York our precious wooden boxes, neatly numbered with white paint and with a list of the contents pinned inside the lids for the benefit of Customs (though all the boxes locked!), were carted across the city from Cunard to Grace Line. Imagine if you tried to do that today. It's a nice example of the changed world. Those boxes could have been packed with high explosives or anything else, for nobody checked. Why would they? Two days later we sailed for Guayaquil, under Statton Bridge, past the Statue of Liberty and Manhattan. Air travel is such an appalling waste. What a loss we have suffered with the demise of the great shipping lines. These days we sit cramped together in a long silver tube, perhaps fretting about blood clots or hi-jackers, transported joylessly through the stratosphere along with our plastic food. …. "35,000 feet, flying time 16 hours 20 minutes, local weather fine but cloudy, please remember your hand baggage, thank you for flying Pan Handle and we hope we never see you again. Goodbye". Now we were skirting Haiti and our first masked booby cruised alongside, very gannet-like in its dazzling white plumage but much lighter and more buoyant, for although it is about as big it weighs only half as much. It is a tropical seabird and doesn't need thick insulating layers of fat, nor the weight to dive deeply and hold heavy muscular fish. At Barranquilla in Columbia the river, in flood, swept thousands of tons of precious topsoil away from the interior to deaden the sea; a double tragedy. Cattle egrets and great white herons perched on masses of floating vegetation and turkey vultures, with huge uptilted wings and red heads, soared above. Ospreys plunged into the swirling brown water and swallows hawked overhead. They were not our familiar birds but velvet-green rough-winged swallows. None of these birds inhabit the jumbo's stratosphere.

We crossed the Equator on December 2nd and Neptune came on board, plastering me with raw eggs, tomato ketchup and whitewash. June got off Scot-free; so much for equal rites. Our first red-footed booby, long-winged and agile, crossed the bows. The little male is less than a third as heavy as a gannet. Later, their tinny, foot-ball- rattle calls sounded day and night around our tent on Tower Island, where thousands nest in the sun-silvered scrub.

Soon Guayaquil 'the hell-hole of the Pacific' drew near and with it a strong and unmistakable sinking feeling. We had been putting off the fraught moment when we would be thrown, as green as Bass mallow, into the turmoil and unknowns of the next, potentially calamitous phase. Somewhere in the Santa Mariana's hold were 11 vital wooden boxes containing all our gear. Would they be put into bond, disappearing into the dark recesses of some cavernous customs' shed, never to emerge again? Fanciful?  Not in the least. It had happened and to far more prestigious expeditions than our tin-pot affair. The august Royal Society of London, no less, had the equipment for its meticulously prepared and officially Brazil-sanctioned expedition to the Amazonian rain forest impounded by Brazilian Customs until it was too late to be used. What if the notoriously unreliable 'Cristobal Carrier', the semi-wreck that plied between Guayaquil and the Enchanted Isles' and just about to depart, would do so before we could extricate our boxes and get them down to the riverside where it wallowed in its muddy berth. There wouldn't be another sailing for weeks, per-haps months, so what would we do in the meantime? What if we were charged heavily for storage and had to wait for lengthy repairs to the Carrier. What if, what if ….. The possibilities were dire with us on such a shoestring and no institutional backing to turn to in an emergency. Strictly on our own, our entire budget and with it the expedition, could so very easily have been wrecked. In retrospect, were we brave or mad.

But somebody, or Somebody looked after us. Roberto Gilbert, a mercurial Ecuadorian surgeon with a British sea captain grandfather, had befriended us on board the ship from New York. He had, as it happened, once removed an appendix from the Chief of Customs. They embraced as brothers and we sped through customs without a box being opened. Even better, Roberto suggested that we store them on the roof of his clinic and arranged for his pick-up to collect and deliver them. This was critical for it had proved impossible to arrange storage in advance for our eminently stealable equipment. This, remember, in 1963. Arranging that sort of thing by letter between the Bass Rock and a Quayaquil shipping agent had simply been beyond us. We had tried, but our letters probably never arrived and even if they did probably posed too many problems, particularly in English. So we were ex-pecting problems, but now everything had been magically solved with a wave of Roberto's scalpel. So we happily installed ourselves in the clean and reasonable Pen-sion Helbig, with three meals and a bed for about a pound a day. We could easily put up with the clinging smell of stale cabbage.

Dr Gilbert was a phenomenon. At the age of 46 he had already paid off the mort-gage on his clinic, worth millions and the best hospital in Ecuador, beautifully built of marble, with departments of radiology, endocrinology, dentistry, etc and fully equipped with kidney machines, lung unit, low temperature gear and much else, which Roberto treated with utter sang-froid. He had performed the first mitral valve heart operation in Ecuador and, with his team had, amazingly, sewed back a sailor's severed hand, although it was later rejected.

Enthusiasm was Roberto's life-style. Soon, not content with mere advice, he plundered his pharmacy for expensive medicines for us to take to the Galapagos. Once, after midnight and still in surgeon's gown, he bundled a wretched intern into the back of his pick-up -"for some fresh air"- and we all roared off on a tour of Quayaquil. Surgeon extraordinaire he may have been; driver he was not. He staggered along in top gear at 10mph and then, as if to atone, began a perfect frenzy of ill-timed gear changes. He wove a crazy route, stopping, turning, stalling and executing split second changes of direction without the least warning. The common folk of Quayaquil knew and loved him as he strode regally through the streets in his surgeon's gown, looking for some potty little article that he thought we should have – like a shotgun. Why he took so much trouble over us I never knew; doubtless June's magic.

On the great day, Dec 11th, we sailed from Quayaquil for a year in the land of magic. Down at the old Cristobal Carrier the scene was animated, not to say agitated. Forty-seven cardboard boxes of provisions from the dark cavern that was Elias Moyorga's, 'ships' chandlers', were ticked aboard. Only 46 came off on Santa Cruz and whoever stole the 47th got our only luxuries – tins of chicken and ham earmarked for birthday treats.. We had to make do for the whole year with tuna, sardines and some quite revolting corned beef hash. For some obscure reason - probably a ploy on Elias' part to off-load some chitterlings (whatever they might be) we ended up with a dozen or more tins of these revolting items. The first one we opened was black, bubbling and stinking. We sank the lot deep in Darwin Bay, attended by our faithful retinue of boobies. Not funny when you've planned for a whole year's food.

We needed a lorry and a gang of labourers to transport our Bass boxes from Roberto's clinic to the Carrier. Our tireless surgeon organised this. Those Ecuadorian porters of indeterminate age and slight though sinewy build, performed prodigies of strength and agility, carrying our devilishly heavy and awkward wooden chests, all sharp corners and unrelenting edges on their bony shoulders, up a slippery gangplank which not only sloped steeply but tilted to one side. It must have been agony. I couldn't have done it, nor could 99% of us softies.

As this steady flow of motley cargo went abroad, an ancient lorry tied together with wire and wheezing chestily, crawled up to the wharf carrying an old rowing boat made of iron wood and weighing the earth. Yet again we owed it to Roberto, who bought it for us 'upriver' for £12. It proved invaluable; our only means of getting to some of the awkward seabird colonies on the horns of Darwin Bay. It went on the 'Carrier's' foredeck along with a few tree trunks and 3 enormous stems of bananas. At the end of the year, we left it with dear old Mr Rambeck, one of the early Norwegian settlers. Doubtless it still parts the waters of Academy Bay.

Our amazing life-saving surgeon still hadn't quite finished with us. Returning, exhausted, to his clinic, we were grabbed and, together with his radiologist, an ex-colonel, rushed across to the Ministry of Defence, where we bumped into the Minister himself just about to set off to the Town Hall Conference Chamber. Our dear Colonel dictated a letter requesting "all civil and military assistance for her Britannic Majesty's' Special Envoy, Senor Dr Bryan Nelson". Then trailing sheepishly behind the Minister, but trying to look like special envoys, we penetrated a succession of armed guards until we reached the chamber in which Government Ministers were actually in conference. There and then we obtained the appropriate signature. Even at this remove the whole things seems Alice in Wonderland, though I have not embellished a single detail. Imagine our Defence Secretary acting like that to help some insignificant, inconvenient, Ecuadorian biologist. He wouldn't get within 10 secretaries of the Minister.

The Cristobal Carrier, on which we were the only 'special class' passengers, with unrestricted access to a large stem of bananas hanging beneath a tattered tarpaulin in the stern, sailed an hour before midnight. What a relief to slip quietly down the muddy Quayas in the tropical blackness, though the vessel inspired little confidence. It was a flat-bottomed landing barge, American war surplus, desperately top heavy as a result of the battered old cricket pavilion which had been tacked on to increase the accommodation. Lacking a keel, it rolled hideously but, as usual in this storm-free zone, the sea was calm and thankfully remained so. A good blow would certainly have finished us off. It is on record that sometimes she missed the islands altogether. Two or three years after our voyage, Roger Perry describes her smoking into Puerto Ayora, goats bleating on deck and on one occasion stern-first because the helmsman could engage only reverse thrust. After this crossing there was to be only one more boat journey before we at last landed on our desert island. How infinitely far from reality had been our daydreams on the Bass. Fortunately our imaginations had been a poor guide, or we would never have begun.

So we relaxed and watched the thousands of seabirds. Waved albatrosses from Hood Island, their only nesting place in the world, soon appeared. At night Madeiran storm petrels fluttered on board, seemingly attracted by the lights. About 150 miles from San Cristobal the sea was alive with them, in strong, wheeling flight

*Christobal Carrier*

28

punctuated by slow butterfly strokes. Then we had our first sight of the rare and lovely swallow-tailed gull, soon to be our constant companion on Tower Island. Flying fish leapt clear of the sunlit sea and a large white-tipped shark cruised alongside. On Tower we became quite blasé about them; in the Galapagos nobody seemed to bother. Fishermen throw over the guts from their catch, then swim ashore amongst them. But the selfsame species around Aldabra in the Indian Ocean is highly dangerous. On that atoll a relatively small fellow, about two or three feet long, took a chunk out of the leg of Tony Diamond (a biologist working on frigatebirds) as he waded across a shallow creek.

At dawn, three days out of Quayaquil, the 'Carrier' slipped into Wreck Bay on San Cristobal, with three raucous blasts of its horn. The gathering light showed an attractive little settlement of sun-bleached wooden houses fringing a pleasant beach backed by cloud-capped hills. It looked idyllic. In 1963 just 1,399 people lived on the whole of San Cristobal, 745 of them in Wreck Bay. Santa Cruz had 609 people, 283 on Isabella, 46 on Floreana and 36 on San Diego. Nearly all the other islands, including 'ours' (Tower and Hood) were uninhabited. Since those days things have changed dramatically, in some ways for the worse, bringing all sorts of problems. This isn't the place to go into the inexorable rise in tourism, the illegal settlements with their desperately unwelcome cats, dogs, pigs and rats, the tribulations besetting the Charles Darwin Research Station and Foundation, the politicking in Ecuador, the rape of marine resources especially sea-cucumbers and sharks (for their fins), the struggle to exclude predatory and illegal foreign fishing vessels, the introduction of potentially devastating plants and insects and much else. It is an on-going and at times desperate struggle. Yet there have been notable successes such as the captive breeding of endangered tortoises and the elimination, after 127 years, of wild pigs from Santiago. Strikes, intimidation, political pressure, commercial greed, the courage of dedicated conservationists – all this is currently being played out in these islands which man, in his usual way, is bidding fair to make anything but enchanted.

When the Carrier wheezed into Academy Bay on Santa Cruz on December 16th 1963, the Research Station was not yet functioning. Several new buildings, well designed and attractively landscaped, awaited the finishing touches but rather little field or laboratory equipment had arrived. The large station warehouse proved an ideal storage place for our precious boxes until we could ferry them out to Tower Island where we were to camp for six or seven months. We used the Station raft to transfer them from the Carrier to the warehouse, though not without severe palpitations, for it rolled alarmingly, threatening to tip our boxes into Academy Bay where they would have sunk like stones. To have lost them so close to the finishing post would have been unbearable. We had no inkling that worse was to come.

Even after Elias Mayorgas's best efforts there still remained a few items such as rope, buckets and soap-powder which it had seemed silly to buy in Guayaquil, so we got them from the village store on Santa Cruz. Wisely, we had ignored advice falsely assuring that we would be able to buy tinned butter and bacon there. We got ours at Selfridges and they proved one of our few treats in a food desert.

During our brief stay on Santa Cruz we were treated to a taste of life in this civilised corner of the Galapagos. David Snow, the Director, was an old colleague from the EGI in Oxford. In fact, it was his excellent study of the blackbird that I was supposed to extend and deepen. I say "supposed" because it didn't work out that way. I never dreamt that our paths would cross again under the most un-Oxford-like conditions imaginable. David kindly took us under his wing until we could arrange our final leg to Tower. He invited us to join him and Miguel Castro his hardy Ecuadorian field assistant, together with a few hens and two cats, on a trip to nearby Plaza Island in Dave Balfour's yacht 'Lucent', which later took us to Tower. At that time the Station did not own a research vessel but was in the process of acquiring 'Beagle II', a splendid old wooden two-masted brigantine which had been dredged up from the bottom of Brixham harbour. Dearly though I love the cold and windy old Bass Rock I had to suppress weak and unworthy comparison with this hot Galapagos sunshine, sparkling blue sea and the turtles, giant mantas, boobies, frigatebirds and swallow-tailed gulls that surrounded us, as we puttered tranquilly along. Plaza itself crawled with sea-lions and land iguanas.

Next day came a visit to William Beebe's "broiling and insufferable" crater on Daphne Major. Actually there are two craters, one immediately behind and below the other which together form a large and deep amphitheatre. The crater walls are steep and bare and the dusty bottoms pancake flat, bereft of even a decent-sized pebble let alone a boulder. Broiling it may be, but obviously not insufferable to the hundreds of blue-footed boobies which nest thickly scattered on this blistering surface, like so many currants in the bottom of a floury bowl. Sheltered from the cooling ocean breeze and roasting in this infernal pit, they seemed to be suffering greatly but maybe they were not so dumb for Beebe was surprised to find Daphne crater so cool! Although completely enclosed he claims that there was more breeze inside than out, the hot air rising whilst cooler air siphoned in through a notch on the rim. Even so, the chicks sweltered and panted in their thick white down but the nights are cold and a chick has to grow its down sometime.

*Land Iguanas*

On the way back to Santa Cruz we passed through a vast shoal of yellow-tailed mullet attended by diving brown pelicans. Thousands of dancing grey phalaropes stippled the surface, frigatebirds idled overhead scissoring their forked tails, boobies dived. Little wonder that naturalists love the Galapagos, but they are only the last-but-one in a long line of visitors, despoilers and intruders to set their inquisitive and acquisitive feet on these enchanted islands, after Incas, adventurers, whalers, settlers and assorted misfits. Now it is well-heeled tourists from Europe and North America, and, ominously, Ecuadorians

*"The boiling and insufferable" crater on Daphne Major, site of a large colony of blue-footed boobies*

*Blue-footed boobies in Daphne's crater*

and others hell bent on reaping a rich if ruinous harvest from the surrounding seas.

In 1964, there were still some of the hardy pioneers who, in the first decades of the twentieth century had left civilisation for the rigours of these islands, amongst them two famous families, the Angermeyers and the Hornemanns. Before leaving for Tower we climbed the five miles of bouldery track now, alas, a surfaced road, from the settlement to the Hornemann's upland farm to pay our respects to this remarkable woman. The track was well used by horses and donkeys; several of their bleached skeletons marked the way like cairns. In 1939 Mrs Hornemann had entertained my Oxford supervisor, David Lack, in this same homely room, drinking no doubt the same delicious coffee from beans which she grew, roasted and ground herself. The black-encrusted coffee grinder certainly looked all of 50 years old. I had always thought of David Lack as I had known him, the archetypal Oxford Don, bookish, fastidious and slightly old-maidish, even a bit of a hypochondriac. Yet he had done his famous work on Darwin's finches out here whilst I was still a small boy. Most visiting scientists had gravitated to the Hornemanns and I confess to a pleasing sense of continuity.

*Brown pelican feeding young*

Up there in the hills the soil is amazingly fertile and this together with the warmth and abundant rain encourages everything to grow – coffee, sugar-cane, tobacco, bananas, mangoes, avocadoes, tomatoes and just about every vegetable. Some people love jungles whilst others, me included, respond more to hills, deserts and small islands. Tropical jungles, humid, encumbered, stinging, pricking, tearing, stumbling and tripping places make huge demands on poise, temper and energy. But when Siegbert Hornemann generously offered to guide us to the highlands of Santa Cruz to see giant tortoises, the opportunity was too good to miss. The narrow, muddy track virtually disappeared in the undergrowth, winding through rhododendron-like shrubs and palo santo trees festooned with lichens. Logs and branches slimy with moss blocked the way. We plodded dourly upwards pushing aside stinging plants and ignoring the fine drizzle until it turned on the tap and soaked us. Water from the trees dripped down our necks and every trip, scratch, snag or sting irritated the more. After two hours of slipping and slithering we emerged into open country at the foot of Table Mountain. Galapagos pintails displayed on a small sphagnum-covered pond; giant tortoises grazed; wild goats browsed around a small crater and a large one-horned bull eyed us malevolently. The return trip was drier and Mrs Hornemann's coffee and cake soon restored us. But my scratches festered and later proved extremely troublesome on Tower Island, where it was impossible to keep the open sores dry even though June piggy-backed me and all my gear across the small creek that we had to negotiate. It would have become a real curse but a French yacht called in soon after our arrival and gave me an injection of antibiotics which quickly and permanently cleared things up. What luck – again!

These pleasant trips soon ended and, after spending Christmas with the kindly Snows, we began loading up Dave Balfour's yacht 'Lucent' for the final leg from Santa Cruz to Tower. Three hundred litres of water and five of paraffin went aboard along with sacks of flour, sugar and rice. Those of our provisions vulnerable to rats had thankfully survived intact in the Station's warehouse. Our wooden chests proved an awkward load for such a small boat with little deck space and poor old 'Lucent' began to look like an overloaded Thames barge.

The big day, our final leg to Tower, was December 27th. This time we really were 'off'. Until then everything had been potentially reversible. Not any more. We weighed anchor at 3 30 pm and motored slowly past Carl Angermayer's lava-block house and out of Academy Bay. Our old paraffin fridge, all the way from the Bass, bobbed along behind in our ironwood dinghy. It all seemed a bit surreal. None of us turned in that

*Dave Balfours yacht "Lucent" taking us to Tower Island*

night. We sat outside in the cockpit chatting quietly. Sandy, Dave's 'crew' was a tough, desiccated little Kiwi, still on his way home nearly 20 years after the 1939-45 war. He had called in at the Galapagos and somehow got no further. We plodded along under power, on dead reckoning. Perversely, now that the big adventure had really begun, I felt no exhilaration. Perhaps we had been too long on the way. Nor, for that matter, did I feel apprehensive although the enormity of our venture really struck home during those pre-dawn hours. A small boat at night on a limitless ocean is a powerful reminder of mortality and in living alone and totally incommunicado on a hostile, uninhabited island for a year we were taking a real risk. June was unusually silent - merely pensive I hoped and not regretful, but it was not the time to ask.

At 7 35 next morning we ran out our anchor chain in Darwin Bay and during breakfast watched apprehensively as the breakers piled across the opening to the tiny landing beach before smashing thunderously against the rusty lava cliffs. This was no idyllic south sea island; not a grass-skirted maiden in sight. Indeed, not even a blade of grass; just dry scrub and dismal lava. Were we really planning to spend seven months here?

It was a long, hot, gruelling day ferrying that awkward cargo ashore from far out in the bay – quite ridiculously far as we later found - and we were more than lucky not to ditch anything whilst transferring those heavy chests from yacht to bobbing dinghy. It was nearly as risky beaching everything in the awkward swell, though they would at least have been recoverable there. For good measure, Dave's ancient Seagull outboard gave up the ghost after one spluttering trip and for the remainder of the day we toiled back and forth with one oar and a piece of driftwood lashed to a pole. Then, just as we had finished and the beach was a chaos of gear, our hearts sank. A scruffy little Ecuadorian fishing boat with seven men on board 'putt-putted' gently into the bay and to add insult to injury anchored nonchalantly within spitting distance of the beach to which we had toiled so strenuously. 'Lucent' sailed out of the bay as soon as the last load was ashore leaving us tired and a bit dispirited amidst the chaos.

If the presence of a few local fishermen seems a trifling matter it did not feel like that. Only once in a lifetime can you sail up to live on your first tropical desert island. It was an experience to be savoured. To have a gang of fishermen muscle in was like a third party on a honeymoon. And they could have proved awkward or even dangerous for all we knew. Fortunately we didn't know how long they planned to stay or we would have been even more dismayed. Weeks later, after they had departed with their cargo of sun-dried fish we went for three months without sight of man, boat or even plane; alone in the world. But on that first night I am ashamed to admit that we harboured deep suspicions. Perhaps it was sheer fatigue, but our gear and provisions piled high on the beach seemed horribly vulnerable in the black night. Every few minutes we peered furtively out of the temporary shelter of our store tent, straining to distinguish between the barely discernible boulders and the creeping shapes of the stealthy thieves, actually sound asleep in the unimaginable bowels of their battered little boat.

*Swallow-tailed gull in flight*

Next day, dog tired after two sleepless nights and a lot of exertion, we struggled to put up our huge tropical canvas ridge tent with poles like telegraph posts and cast-iron pegs that would have anchored the 'Queen Mary'. It seems crazy to cart such things half way round the world but we would have had great difficulty buying equivalents in Ecuador, let alone on Santa Cruz. Maybe we imagined gales in which the normal weedy tent would have been useless. It took us half a day to raise and guy it but when it was done we had a home beyond imagination. Behind us we looked out onto a lagoon fringed by nesting frigatebirds, masked and red-footed boobies and swallow-tailed gulls. Outside our front flap a white, coral beach opened onto the wide blue circle of Darwin Bay. Galapagos pintails

The white form (black tailed) of the red footed booby on Tower Island, Galapagos, where commonest form is brown

A brown form of the red footed booby paired with a white form, showing that mixed pairings occur

A juvenile red footed booby demonstrates the famous tameness of Galapagos birds. Here on June's hand

*The camp on Tower Island*

visited the lagoon and night herons stalked the edges like evil gnomes. Wandering tatlers played tag with the tide just like our own sanderling. For garden birds we had Darwin's finches, mocking birds, yellow warblers and Galapagos doves. The Bass had been climaxed.

We knew that there was no water on the island so we took the materials to make a solar still and to our amazement it worked. Although by no means original, it wasn't a commercial model either; the prototype had been tested on the Air Ministry roof in London. It was simplicity itself – just a shallow rectangular pit in the sand lined with black polythene and filled with seawater. Over it we built a wooden frame like a greenhouse roof to support a tough, clear plastic sheet of a material called 'mylar' (kindly supplied by Du Pont), which we bent under at the base of each sloping side to form a shallow trough. The two ends of the greenhouse were sealed with more black polythene and the whole thing set on a slight tilt on the long axis. The sun evaporated the seawater which then condensed on the underside of the mylar, trickled down into the troughs and thence into sunken collecting cans. Eureka, it worked. The snag lay not in the design but in the tiny crabs that punctured the black lining and allowed the seawater to drain into the sand. Evidently there had been no crabs on the Air Ministry roof. We had to unpick one end so that June could wriggle in to try to fix it. Alas, I have lost the photo of my naked wife inside her transparent case. At best we managed five or six pints a day which might have been just enough for survival. Black rubber sheeting would have been more durable than the plastic.

We had barely settled in before a coal-black finch with a thick conical bill hopped boldly into the tent. He was one of the famous ground finches whose ancestors had (eventually) steered Charles Darwin's thoughts towards the theory of evolution. Straight away he perched on our fingers, a charming tameness that has cost

*The solar still*

thousands of Galapagos creatures their lives. He greedily sipped water from a spoon and adored almonds taking fragments from our lips with the utmost gentleness. Probably our tent was outside his territory; for he always slipped in surreptitiously and only when safely back on his cliffside cactus did he burst into his simple song. Every morning he wakened us by delicately nipping a nose to 'ask' for his almond. When you think about it, that is quite remarkable. How did he know that it would work – what made him connect that act with subsequent gratification and what made him do it gently rather than forcefully? Think how violently some birds importune their parents for food. We left Tower in July and returned in November and who should fly down to the beach to greet us but our little black finch. It was a wretched moment for we had absolutely nothing to give him.

Finches, doves and mocking birds were our constant companions. The clownish cinnamon-coloured doves with their foolish cerulean-blue eye-rings pattered brazenly around the tent before abruptly shooting off with that decisive clatter of wings. The buccaneering mockers were an altogether tougher proposition, driving a normally composed June to dancing tears of frustration. They defaecated into the flour, fell into any open jar including one half full of vinegar and tugged dementedly at anything loose. We stuffed socks and rags into every conceivable entry to the tent but the little devils simply hoisted them out like so much leaf litter and after beady-eyed inspection, head cocked, began their thorough and destructive tour of inspection. They were merely being their usual opportunistic selves, essential for survival in the hostile Galapagos, but it is not always easy to take such an objective view of irritating behaviour especially when one sprinted off with our last slice of bacon.

*Finch on a spoon*

Every day we were awakened by the welcome sun filtering through the green canvas and illuminating the untidy clutter within. Much of our year's food supply lived inside with us because the separate fridge tent was too small; it was made of ordinary light canvas which soon rotted in the fierce sun until even an alighting

*mocking
birds in bowl*

dove punctured it. A pair of swallow-tailed gulls nested cosily against one wall of our living tent, inches from my head. It woke us every morning with gutteral croaks and weird glottal clicks. Red-footed boobies beginning a day's courtship before it got too hot, flew in with tinny, rapidly accelerating chains of sound. The hoarse 'yakking' of a frigatebird beating heavily towards a fully-grown youngster cater-wauling like a stuck pig. A short burst of song from our friendly finch in the cactus. Everyday Galapagos sounds; there was no sleeping through it even if we had wanted to and in any case the tyranny of the spring balance, Vernier calipers and measuring rod (a metre long to cope with frigatebird wings) immediately asserted itself. The daily rounds had to be done before breakfast. As it had been on the Bass, so here, there was simply no substitute for living with the birds.

A deep sea-washed crevice provided a clean and effective loo, so long as you didn't fall into it. Given the right light, you could peer down and watch the gaudy parrot fish baring their ludicrous dentures as they browsed on the waste,. A quick dip in the sea with the warm sun as a towel and then breakfast, the best meal of the day. Oatmeal with reconstituted milk fresh from the fridge, bread baked in our old flat-wick oven, tinned butter and marmalade. The mockers loved breakfast time and given the ghost of a chance flew up and snatched bread from our hands. Apart from pathetic attempts to sweep out the tent using an old toothbrush there was little housework to do, other than the odd foray after a giant centipede that, obeying its thigmotactic, geocentric and antiphotic tendencies had crawled into the tent and squeezed into any old crack. They were terrifying beasts, trailing their armour-plated bodies raspingly across the floor, claws scratching and formidable jaws at the ready. The largest measured 11inches and broad to match. Even with the head chopped off they ran like express Japanese trains. I think the fishermen's catch dry-

*The lovely swallow-tailed gull is territorially aggressive*

*The cryptic chick of the swallow-tailed gull, preyed on by owls and frigatebirds*

ing on the beach attracted them but I wouldn't have been surprised to see one making off with a booby chick.

Even several days after our landing on Tower we were still piqued by the mere presence of 'San Marco' and her industrious crew. There they were, virtually in our front garden, interested in our every move, constantly ashore and inevitably, given the tiny beach, right in front of our tent to collect wood for the brazier which burned on a cradle overhanging the stern. If the boat had caught fire we could never have

*Great frigatebirds feed their free-flying offspring for several months*

fed and watered seven men for long and none of us had any way of contacting the outside world. Their presence was all the more galling because they had far more right to be there than we had. Seven men, hardy and industrious, worked and slept on a battered little boat scarcely more than seven metres overall. Amidships stood a 50 gallon drum of seawater in which they washed their filleted catch before laying it out on the coral beach to dry. Each grey dawn they puttered out from the anchorage trailing an escort of boobies, to fish in the open sea on the far side of Tower Island. Occasionally, when checking the masked booby colony on that side, we saw them tossing up and down in the chop, fishing endlessly. They returned in mid-afternoon, to gut, salt and dry their catch, another long and tedious chore. Then a frugal meal and nothing to do but retire below deck in a squalor we could only imagine. Then the next grey dawn. They were a swarthy, ragged, villainous-looking crew, but as pleasant and courteous as one could wish. They never came for treatment for cuts, boils or toothache without a 'thank-you' fish. They toiled for six weeks and then, hoisting a filthy rag of a sail, set off back to Wreck Bay, navigating, if that's the word, without a compass and relying in the event of engine failure on the efficacy of match-stick crosses dropped overboard. Roger Perry records that a small local fishing boat with four men on board was several days overdue at San Cristobal after a passage that should have taken 10 or 12 hours. They carried no dinghy, sail, water, food or compass and their engine was known to be unreliable. Two months later an American tuna boat arrived at Academy Bay carrying the missing boat which they found off Cocos Island. Nothing was ever discovered about the men.

A great and rare delight of life on a desert island is the freedom from intrusion – no unwanted visitors, telephone calls or other unwelcome distractions. For us, of course, it was a totally artificial freedom because it depended entirely on outside help and on the food we took with us, but as a temporary retreat it was wonderful.

*Pair of masked boobies and seven week old chick*

*A fracas between a masked booby and a juvenile great frigatebird (Galapagos)*

The sense of detachment that came with the isolation was a curious feeling, and completely different from anything you can get from a brief holiday, however remote. A couple of weeks is nothing, but as the weeks stretched into months without sight of man or his artefacts we felt more and more self-contained. I am not by nature gregarious and for me the whole venture would have been ruined had it involved other people with all the petty tensions that are almost inevitable under such restricted and challenging conditions. If we had been part of a normal expedition I would undoubtedly have found it intolerable whereas with just the two of us the days, weeks and months slipped peacefully past with never a cross word. Some people find this harmony hard to believe and a few misguided souls think it not only downright unnatural, but even undesirable. Each to his own, I suppose, but it certainly made for pleasant co-existence. Equally, though, I would never have contemplated such a venture on my own or even with a male companion to share the

*Lava heron and*
*Pintail*

chores. Our particular circumstances were ideal; the three years on the Bass had been admirable preparation, we were young (ish!) and in love.

Billions of people face an unbelievably dreary future, bleak, forbidding and deprived – hardly even survivable. We seemed to have hit the jackpot. In the evenings, our old Tilley lamp shedding a pool of mellow light against the black tropical night, we daydreamed amidst the comfortless clutter of our crowded tent. We were nearly penniless, without home, job or possessions, but the future seemed wonderful. A trivial but still exciting prospect was getting a landrover. I had always coveted one and in 1965, split new and with a hard top, it cost precisely £640. That landrover took us through the loose sand of Wadi Rum, just as it was in Lawrence's day, long before there was a road. It forged through miles of flash-flooded desert with water halfway up the doors, across miles of spine-jarring black basalt boulders shimmering in such heat that even desert-adapted larks gaped. It survived sea-crossings, impoundment and stoning; carried pi-dogs, jackals and a wolf cub and two very senior British Royals. For short periods we lived in it, short-wheel base though it was. It ended its days on an Aberdeenshire farm or perhaps, like nearly three quarters of landrovers ever built (or so I'm told) may still be going strong.

For much of our time on Tower we wore nothing around the camp and when we did dress it was mainly for protection against thorny scrub. It was such a treat to swim in a warm sea, dry in the hot sun and go about our chores. I think, if they tried it, most people would quickly lose their prejudice and under suitable conditions find nakedness natural and pleasant. We really are rather silly, condemning innocent nudity but avidly watching grossly distasteful films and TV. But then most Europeans, especially Brits, look best fully clothed. Pale flab is not pretty.

After a couple of months on Tower we felt pretty relaxed. We had established a work routine and results were beginning to look interesting. By this time we often cooked supper outside under the velvety black sky. The sea surged rhythmically against the coral, soothing and only half heard. Great saturnid moths as big as bats came to our light, giant centipedes foraged like the predators they are. We baked bread in a biscuit tin packed with fragments of coral to spread the heat. That homely smell of new bread seemed faintly incongruous on a tropical desert island. To go with the bread we cooked delicious crayfish for which we dived offshore; they were only around nine metres deep. Once, June baked a chocolate cake. Then we danced on the sand to fuzzy music from Radio Belize, picked up on our little transistor radio. I write all this, accurate in every detail, yet am quite unable to recapture even a glimmer of the feelings that went with it. That, to me, is a minor tragedy for I would love to re-live those magical moments. It is a lucky person who can recapture the intensity of pleasurable experiences. We recall the facts but not the feelings (though electrical stimulation of the appropriate part of the brain might do the trick!). Sadly, it is probably easier to re-live fear than pleasure because it is etched more deeply. I imagine the victims of torture find it all too easy to re-live the pain and terror. Torture is surely one of man's most hideous practices, setting him apart from all other animals in its conscious intention to cause agony. A cat playing with a mouse is a completely false analogy.

Early in February a few sweet showers transformed Tower Island, throwing a delicate green veil over the bleached skeletons of the trees, which had looked drift-wood-dead. For a brief fortnight the black and rusty brown of the lava was softened by this evanescent foliage and the air became deliciously scented. Then the leaves withered and Tower became a sun-baked lava desert again. It took real effort to cross the ankle-twisting lava to the far side of the island laden with cine-camera, telephoto lens, heavy wooden tripod, binoculars and all our gear for ringing, meas-uring and weighing boobies. By the time we had shot 400m of film, caught, ringed, weighed and measured 15 stoutly resisting adult boobies and twice that number of chicks, then humped all the gear back again, we were more than ready to jump into the lagoon and drink half a gallon of liquid.

By late February the island was vibrantly alive. The rain had stirred up the land birds, the boobies and frigates were courting madly, the night herons building and the small black marine iguanas had begun to dig burrows and lay eggs. More than a thousand had gathered on one small beach, motionless as lumps of coal. But in the tropics there is no guarantee that things will proceed without let or hindrance. Dur-ing nearly half a century I have not known a single year in which the gannets of the Bass Rock have had their breeding cycle disrupted or even much delayed by poor weather or shortage of food. It is far different in the tropics. On Tower Island the nesting activities of the boobies suddenly ground to a halt because there were no fish, or at least few that the boobies could get at. This can happen as a result of abrupt changes in ocean temperature, even though these may be short-lived. Shortly afterwards, and just as suddenly, everything picked up again. This was all new and exciting to me, accustomed as I was to the predictable seasons of north temperate seabirds.

*Marine Iguanas on Tower Island*

*Marine and
land iguanas*

So the weeks slipped by, until on March 5th 'Lucent' returned with 250 litres of paraffin, 200 litres of water and a heap of fresh fruit and vegetables from the wonderfully kind Angermeyers. After cheerfully helping to drink the beer which they brought they departed in good fettle leaving us to our solitude once more. To our very real satisfaction we felt not the slightest urge to go with them. Stimulated by this 'relief' we set to work and tidied the camp. Several weeks-worth of empty tins were pierced and consigned to the abyss of Darwin Bay. Ever curious, the red-footed boobies trooped out to supervise.

March 14th was my 32nd birthday, time for a treat and a bit of exploration in our old wooden rowing boat. As we crossed the bay an enormous manta ray glided directly beneath us its wing tips sticking far out on each side of the boat. It could easily have carried us on its back or overturned us with one galvanic flap. Of course it did no such thing and we simply admired its grandeur. Several white-tipped sharks accompanied us, graceful predators but, in the Galapagos (though not elsewhere) posing no danger even to swimmers. A frigatebird repeatedly picked up and then dropped a fledgling shearwater which, I supposed, had attracted its attention by some slight sign of weakness. Predators are astonishingly quick to notice any peculiarity, just as a lion or cheetah on the Serengeti plain picks out a marginally unfit

*Fur seals resting*

gazelle from hundreds of fleeing animals. A herd of dolphins fed in water-churning surges at the mouth of the bay which we avoided like the plague because of the strong current that would have swept us to certain and horrible death in the open ocean. A few fur seals hauled out at the base of the cliffs.

At one point in the cliffs fringing Darwin Bay we found a negotiable crack, new to us, which later became known as 'Prince Philip's Steps'. I'm not sure who first discovered them – possibly it was Miguel Castro, David Snow's resourceful field assistant - for David used them before we did. But HRH it certainly was not. Still, 'Castro's Steps' somehow lacks the panache of 'Prince Philip's Steps'. The crack

*Short-eared owl
and storm petrels*

emerged onto a great plain of volcanic ash and shattered lava crust with a few scattered boulders and stunted trees, a typically hostile Galapagos scene. Further round the north horn of the bay we discovered a vast field of weathered lava, utterly bare and desolate, a broken, wafer-thin crust with shattered scoriae piled haphazardly like millions of smashed plates. Above this, thousands of storm petrels flitted in mazy flight.

As Atkinson wrote about Scottish petrels "the flight seemed feather-light, to have a buoyant, butterfly aimlessness". The air was thick with them but the only sounds in an eerie hush were the slight scratchings and the muffled rasp of wings as, every few seconds, a petrel landed and slipped quickly into a crevice or emerged like a conjuring trick from beneath the lava. Before entering a crack they pattered over the surface with dangling feet, just as they do over the sea. Later, David and I collected some of the hundreds of wings lying in heaps in the 'owls' parlours', parted from their owners by the predatory Galapagos short-eared owl and discovered that they belonged to two species of petrel, the Madeiran and the Galapagos stormy. Nobody knows why petrels engage in these mass flights, though it has been shown that on some Scottish islands where large numbers of our own stormy petrels fly around at night most of them are non-breeders, lacking the brood patch of breeding birds. Catching and ringing hundreds of them has shown that they may visit different islands on successive nights though why they do so is a mystery which it will be more than difficult to resolve. One can think of possible reasons such as assessing whether there are potential mates and/or suitable nest sites but proving it is another matter. Simple questions, complicated answers.

By early April the ground finches were feeding free-flying young. Each male tended a single juvenile which stuck to him like a shadow, begging incessantly. At intervals he regurgitated food piece by piece with a quick jerk of the head. The juvenile's lower mandible was conspicuously cream-coloured which I suppose acts as a visual marker in the gloomy thickets of cactus where they live.

So the months on Tower elapsed and in July it was suddenly time to pack up and transfer, lock, stock and barrel, to Hood Island, across the Equator in the south of the Group. We had arranged the date early in the year for we had no contact with Santa Cruz and the Research Station. It took us three whole days to coax our shambles back into the old green boxes, dismantle the tents and re-crate the paraffin fridge. When we had finished and the low shingle ridge on which we had lived for seven months was once again serenely clear, trusty old 'Beagle' which by now had begun its service as the Station's research vessel slipped into the bay, trailing its

*"Beagle II" the research vessel of the Charles Darwin Station, which took us and our team across the equator three times in 1964*

faithful retinue of boobies. She dropped anchor and cleared decks to receive our motley gear. This time the process was far less traumatic than when we had landed from 'Lucent'. Beagle was well equipped with dinghies and outboards and the skipper, Carl Angermeyer, had an able and willing crew which included Julian Fitter and Richard Foster. Julian eventually married an Angermeyer, returned to Britain to fit out a boat of his own and then sailed back to the Galapagos with wife and baby. Then he used his boat for charters in the islands. After several years Julian returned to a business life in London, the very fate from which he had fled in favour of 'the good life'. Carl, his crew and their pet marmoset entertained us royally on Beagle and Carl expertly trimmed my shaggy locks and beard. After months of tinned and dried food the fresh turtle was tender and delicious but our consciences pricked sorely even as we licked our lips. Marine turtles face more than enough threats. Tourism and beach developments ruin their breeding places and discarded fish nets and long lines ensnare thousands in the open oceans. Untold numbers have been killed in recent decades. Yet another black mark for us.

At 17 30 hours, July 10th 1964 we left Tower after 194 days, our notebooks full of hard won data and heads well stocked with ideas. During the night the helmsman, whose name shall be shrouded in decent obscurity, snoozed peacefully at the wheel allowing the strong current to throw us off course so that by morning we were heading back to Tower, by then only 40 kilometres away. Carl was not amused but he was a kindly fellow and nobody was keel-hauled. Sadly, in the 1990s. this handsome, gifted, flamboyant and seemingly indestructible man suffered a stroke. But what a life he had led after he and his brothers rejected Hitler's Germany in the 1930s and sought their fortunes in the enchanted isles.

The day after our return to Academy Bay Carl invited us for a meal. Inside his lava block house the furniture gleamed, the tableware was elegant and the squalor and hardship of his first years on the island with his brothers Fritz and Gus must have seemed a distant memory. For us, though, his hospitality was a brief treat before we set sail for Hood, our second Galapagos desert island, with its blue-footed boobies and waved albatrosses.

*Mocking bird in a*
*vinegar jar*

*Waved albatross*

## Chapter 3

# HOOD ISLAND: SEA-LIONS, GOONIES AND OTHERS

*"I was on the level and when he came open-mouthed at me out of the water, as quick and fierce as the most angry dog, I stuck the point into his breast and wounded him all three times he made at me, which forced him to retire with an ugly noise, snarling and showing his long teeth at me out of the water."*

God spewed lava on Tower and rained stones on Hood Island or Espanola. Punta Suarez where we camped was all stones; not the wickedly sharp and brittle lava sheets of Tower Island but smooth, age-worn reddish boulders. On the south side of the point the rusty cliffs fell onto a deeply-fissured terrace pierced with blowholes which shot out booming spumes of spray. On the opposite side, where we camped, a delightful sandy beach shelved away into cool, clear water. Behind the beach, stretching towards the centre of the island, silvery scrub and cacti formed, in many places, an impenetrable tangle.

We arrived in good old 'Beagle' after twelve hours sailing from Academy Bay on Santa Cruz and with the usual willing help and banter from the crew we had everything off-loaded and the tent up within a couple of hours – a far different job than when June and I struggled to raise it for the first time on Tower. This time our island stint was to be about four months and although, like Tower, our island was arid, waterless and uninhabited, we expected our 600 litres to last comfortably. David Snow, our kind friend and the Director of the CDRS had just been succeeded by Roger Perry, until recently a BBC man, who had applied for the job on a whim and at the last possible moment, not really expecting to get it. As sometimes happens, the slim-chance applicant who at first sight might have seemed a rather odd choice, for he was in no sense a biologist, turned out to be just what was needed. He proved to be shrewd, diplomatic, insightful and a good, trustworthy friend to those who worked with and for him, though no push-over for anybody who tried to steam-roller him. At that time the administrative side of the Station was, not surprisingly, a bit amateurish for the first two Directors were scientists, trying to establish sound projects and concentrating

*Hood Island camp*

on getting the fledgling station up and running, more than enough to keep them busy. Roger himself became a hardy and adventurous explorer of some of the wilder reaches of the Galapagos Islands. I well remember him on his first trip to Hood Island still in his BBC suede shoes, bow tie and pork-pie hat. His account of his six years as Director ('Island Days', 2000, Minerva Press) is by far the best description of the early days of the station and of some Galapagos characters.

When we landed the beach, thronged with sea-lions, stank of droppings. Heavy bodies had polished the boulders like glass. Although innocently ignorant of the fact, we were setting up camp in the midst of breeding sea-lions. All day and every day, and most of the night, sounded the monotonous blubbing coughs and roars of bulls patrolling offshore. A glance from the tent showed sea-lions surfing and cavorting not merely on top of but entombed within the glassy, towering walls of thunderously breaking rollers along which they shot like torpedoes. At night heavy breathing and snoring betrayed sea-lions snoozing hard up against our tent. Heart-stopping roars and crashes betrayed the pell-mell battles of rampaging bulls. More than half-a-ton of sea-lion tripping over the tent guy-ropes sometimes threatened to pull everything down around our ears. One night, goaded too far, I rushed out and belaboured a be-mused bull with the first thing I could lay hands on, which happened to be the aluminium legs of a folding table. They bent like soft wax. The morning after a night in which a huge bull had roared heart-stoppingly just outside the tent I thought of what might have happened if the beach-master had cornered him there. A ton of battling sea-lions would have crashed down on top of us. So I built a barricade of sorts, but it was a mere token, to be brushed aside like candyfloss. From then on, our campsite, which had never been eligible for a boy-scout's badge, became a veritable tip and after

we shot a goat and hung the joints under the veranda it became a cross between a rubbish dump and an abattoir. But after weeks of weevilly spaghetti and sardines that
goat was Michelin 5-star, indistinguishable from lamb.

The focus of the beach-master's interest, his consuming passion, was the harem of
sleek cows hauled out, snuff-dry, on the warm sand, well beyond the reach of the curling, ice-green breakers toppling lazily and thunderously to seaward. To fend off intruding bulls the beach-master patrolled ceaselessly just offshore, backwards and
forwards in a series of sinuous, porpoising dives. Each time he surfaced he roared, a
hoarse bellow that followed a heavy inhalation like a donkey gathering itself for a massive "bray". There were plenty of intruders for as in all mammals which are hugely
polygamous but in which the sex-ratio at birth is more or less equal, the reigning
monarch is constantly fighting off other sexually mature males. Eventually these cov-
etous bulls are compelled to make a serious bid for the harem but they have to choose
the right time. The most propitious moment for the challenger comes when the old one
is exhausted by fasting and mating but this was not easily judged. Usually the beach-
master saw off his rival with a single galvanic charge, shoving walls of solid water
aside like foam. Often they clashed in a maelstrom of churning water before disengag-
ing. Then came the pursuit, two humped, glistening black backs hurtling out to sea.

*Sea-lions and iguanas*

We saw a few successful takeovers and knew that others had happened because we could recognise some of the beach-masters by their scars. Our favourite old tyrant, for some obscure reason called George, probably after the principal keeper on the Bass, had a piece bitten out of his forehead. One morning, a new bull was patrolling, probably somewhat younger, with a flatter forehead than that of the impressively domed master. Around midday George returned and the new bull slipped furtively away to sea but a few hours later he came back again and roared whilst George was hauled out, apparently asleep. He awoke and immediately roared repeatedly, utterly silencing the newcomer, who kept low in the water before departing in a series of casual dives. But soon afterwards George himself disappeared, presumably succumbing to a new challenger unseen by us. That sort of thing is so often the snag with fieldwork – something significant happens but you can't know what. Still, painstaking observation over a long period can eventually put together a reasonably complete picture, as studies like Diane Fossey's on gorillas, Jane Goodall's on chimpanzees, and Norton-Griffiths' on elephants have shown.

The sea-lion take-overs were quick affairs rather than desperate battles but all the same they had an electric menace, a bruising clash of heavyweights. On one occasion, around midday, a huge beach-master rolled in the shallows when an equally impressive bull approached from one side, reared up and gazed intently at him. The master soon looked up and moved menacingly towards him, meeting an equally threatening approach from his rival. They clashed and after a brief fight the new bull, distinguishable by a long, narrow wound on the back of his neck, chased the other out to sea before returning to the beach. Yet the very next day the patrolling bull lacked the neck scar. The day after that, the scarless bull was himself displaced by a light coloured specimen with a circular neck wound. A mere fifteen minutes later along came old narrow neck scar again and after a short, sharp struggle took over the beach once more. The next day, lo and behold, he had disappeared only to pop up the day after and dispel the latest hopeful. So it seemed that there were several bulls attempting to take over the beach and one, dominant for a while, able to go away, perhaps to feed, return and resume command.

Despite the fierce clashes we only once saw a badly mauled challenger. He had a large, raw wound on his back which looked nasty but in an animal of such enormous vitality may not have been dangerous. Nevertheless, the following day he was lying amongst the bushes and fled precipitately when the master bull approached. Further along the beach another three hoary bulls carrying slight wounds, were hauled out. When dry they look much fiercer, with quite savage faces. The exhausting fights, constant patrolling and frequent mating must be hugely taxing, and we found the remains of several enormous bulls which we imagined had been exhausted, utterly spent, and ready to die. But who knows?  If a bull impregnates many females in a short reproductive life, it can be as effective as living longer and breeding more slowly. Obviously such a system can work only if, as in the sea-lion, the male contributes nothing to the actual rearing of his offspring. The red deer stag can do it but the male wolf, who hunts to feed his litter, cannot. Nor could early man, who was also a hunter.

On Punta Suarez the cows dropped their pups amongst a jumble of boulders; the beach was used solely for courting, mating and resting. The bull was a ponderous suitor and the sleek cow led him a merry dance. He recognised an oestrous female by scent but she did not allow him to mount straightaway. She eluded him and gambolled blithely along the beach with the fat male in clumsy pursuit like something out of Punch. Then they usually played, the male nuzzling her, particularly in the armpits, a ticklish spot which made her jump. She reared up and bit the skin of his chest and throat whilst he kept up a muted version of his territorial bellowing. The whole affair seemed thoroughly amiable. When finally she allowed him to mount, usually in shallow water, he buried her beneath his enormous bulk. It may have been the sea-lions that broke the camel's back, for on August 12th 1964 June announced, with uncharacteristic vehemence, that the old bull got on her nerves, she'd had enough of the daily grind and the privations and wished she were home! I was only amazed that she had stuck it for so long without a word

*Hawk*

of complaint. Truly a girl in a million. Nearly 50 years on, she still looks much the same, but has changed inside!

Several sea-lion pups were born in August. The afterbirths attracted lots of scavengers including lizards, mocking birds and Galapagos hawks. One pup born not far from our tent became positively affectionate towards us. A second approached us but its disapproving mother siezed it by the scruff of its neck and lolloped into the sea with it, swimming around with the pup below the surface most of the time. Only occasionally did its tiny head break surface and we were sure it would drown, but when at long last she did bring it ashore it managed to crawl weakly up to her. Even large pups, with well-pregnant mothers, continued to suckle. These boisterous youngsters were the bane of the adults' life, rollicking ashore, wet and full of bonhomie, and bulldozing their disruptive way into the ranks of their dry, blissfully snoozing fellows, crawling over them, flopping onto them and nuzzling them until all was turmoil.

Fascinating though they were, the sea-lions had to take a back seat. It was with the albatrosses that we really hit the jackpot, no small stroke of luck for we hadn't even tried to time our visit to coincide with their courtship – virtually impossible anyway – and could easily have completely missed all the action. But they fell over

*Waved albatross in flight*

themselves to co-operate and we were treated to the full gamut of their extraordinary display as well as their chick rearing behaviour. Like the fur seals and Galapagos penguins the albatrosses had reached the unlikely haven of these tropical islands via the cold water of the massive Humboldt current which carves its way north from Antarctica. Although less spectacular than the majestic royal and wandering albatrosses they are still pretty impressive, with the dark, gentle albatross eye, a great hooked-and-plated yellow bill and the finely vermiculated breast which gives them their name. Many of them had downy chicks the colour of dark chocolate, delightful when small and clean but spectacularly ugly when large and bloated with the oil which the adult regurgitates. Their beautiful soft down soon becomes soiled and spiky, like a punk hair-do. We crept up to watch at close quarters how the adult transfers its cargo of oil and squid to its eager offspring, as much as 500g in less than five minutes. It is easy enough, if a trifle undignified, for a booby to pass on the fruits of its fishing: the chick simply sticks its head as far down the adult's throat as its anatomy allows and grabs the bolus of fish. It looks uncomfortable for the adult but it is effective and usually thwarts the piratical frigatebird, which in full flight is capable of surgically splitting parent and chick asunder and grabbing the fish. The albatross does things quite differently. Whilst gathering food at sea it manufactures oil in part of its stomach and flies back to base with a tanker-load. The chick places its lower mandible across the lower half of the adult's open bill, at right angles, and the parent then squirts oil up from its stomach. It emerges in controlled jets, straight into the trough of the chick's bill and is rhythmically gulped down. In this way a steady flow passes from parent to offspring without any waste and with precious little chance for anybody to steal it en route. The chick's stomach swells visibly as the oil hoses in, a bit like a flat tick turns into a fat grey ball as it sucks blood. Well-grown albatross chicks are abysmally ugly. Their great stomachs sag over their soiled webs and their down becomes spiky with oil.

Brown eyes gaze stoically out of a ridiculously coiffured head. After a gargantuan feed it is an effort to waddle slowly into the shade of the nearby scrub. They look as though they could do with a pair of hands to support their bulging stomachs. The part grown chick weighs more than 5kg and is simply unable to stand upright for long. It has to squat on its tarsi with its belly bulging forwards and ideally should nest on a smooth velvety sward with a few friendly bushes for shade. Then it could settle its paunch gently into a convenient hollow and snooze blissfully until the next feed. Instead, on Hood, it had boulders to trip over, innumerable and inescapable, and thorny scrub to entangle it like a sheep in a briar patch. Along Hood's north coast they actually chose to nest amongst dense scrub through which the adults threaded a torturous course, in some places for at least 200 metres from their cliff-edge landing place. Yet there were plenty of lovely clearings fringed with bushes for cover and perfect for landing, but devoid of albatrosses. Odd.

Sometimes chicks attacked by adults stumbled and fell whilst trying to escape, especially dangerous near the cliff-edge because they lacked the instinctive fear of genuinely cliff-adapted seabirds. Once, walking at some distance past a large chick, we were appalled when it shuffled to the brink, stood uncertainly for a moment whilst we froze, and then simply stepped over. Perhaps it is the absence of a fear of cliffs which makes the apparently ideal strips of ground near the edge unsuitable for nesting. Waved albatrosses are restricted to Hood Island so it isn't possible to see what sort of

*Mocking bird with albatross and chick*

*Young albatross*

habitat they choose on other islands, which might have given some clues.

Many of the Punta Suarez albatrosses were busy with chicks but some, still dis-playing may well have been pre-breeders, for paired albatrosses need a year together before they even attempt to breed. These care-free birds gathered in groups beneath the sun-silvered scrub, emerging in the mornings or late afternoons to dance with bill-clattering, loud whoops and demonic 'ha-ha-has'. They walked with a comical ritualised gait, head swaying in counter balance from side to side and great webs lifting high.

Animal display is endlessly fascinating. In some, for example the great-crested newt, the male alone performs the display whilst the female merely watches passively whilst he goes through a complicated ritual, rather like a washing machine proceeds from fill, to wash, rinse and spin. All the machine needs is a constant supply of energy and then it goes through its pre-set routine. Similarly, all the male newt needs is the presence of an interested female and he will display, one component predictably following another. By contrast there are displays such as the gannet's greeting ceremony in which both partners play equal parts, and indeed perform virtually identical actions, like two washing machines working synchronously. The albatross dance,

*Albatross displays*

though, was like two washing machines, each with the same repertoire but working out-of-phase with each other. Yet not randomly out-of-phase. The 'spin' of washer 'B' was more likely than any other part of its repertoire to follow the 'rinse' of washer 'A'. The big thing about such behaviour is its perfect predictability. There is simply no question of the albatross, or the gannet, or the newt organising things differently.

To record behaviour we used notebooks, simple shorthand and a stopwatch. The dancing was fast, furious and vocal, each bird snapping into and out of its various postures with whinnying calls, bill-clappering and loud "clunks" as the massive bills snapped shut. We caught and ringed some birds and found that groups remained together for several days although individuals might drop in and stay for a while before leaving, a bit like a hippie commune Most of the pairs which formed and danced during these 'shore-leave' periods were not even semi-permanent, which suggests that the object was not the formation of a durable pair-bond but rather the perfecting of this complex dance, which plays a vital role in their eventual life-long partnership. Sometimes, indeed, things did go awry like partners getting muddled in say 'The Duke of Perth' at a Scottish ceilidh and on such occasions the male broke off in an apparent tantrum and even briefly attacked his partner or re-directed his frustration by attacking a nearby male. Our crying need, of course, was to know the age and status of the dancers but, alas, this information was beyond our reach. Later, Mike Harris, another seabird researcher, showed that these albatrosses do indeed require at least one season's display as a pair, before breeding for the first time. After that they usually stay together for life; the birds that bray together stay together. That is the function of dancing, to help partners form a strong bond before they embark on the long and demanding process of breeding. Might it be a good idea if we treated courtship similarly?

Like most seabirds, breeding albatrosses make heavy demands on their partner. Day after day, even week after week, the bird on duty must sit on the egg without food, water or sight or sound of its absent mate. For a tropical bird such as the waved albatross it is hot, tiring and, for all we know, mind-blowingly tedious though we can have no idea what, if anything, can bore an albatross. Endless quartering of the waves would seem pretty dull! So, for efficient co-operation, the pair-bond has to be strong and the partners have to be compatible. One might not imagine that seabird partners could be incompatible, but they can, especially in the important matter of sharing the work. In kittiwakes, for instance, one partner (usually the male) may fail to shoulder his fair share of incubation and the care of the young. Such pairs tend to break up. The long period of albatross display may be one way of weeding out incompatibility. Once they are firmly bonded they remain together for life and this, moreover, with precious little subsequent display. They have nothing equivalent to the gannet's frequent and ecstatic greeting ceremony. So it is well worth while getting it right in the first place. Many long-lived seabirds form 'trial' pairs in so-called 'clubs' before settling down. Prolonged interaction must tell them something about each other, though precisely how it works is a mystery and likely to remain so.

Many of the albatrosses lost their eggs because for some reason which we could not fathom they frequently nudged it around using their closed bill and sometimes it fell into a crack. Naturally they lacked the intelligence to lift or lever it out and so it just stayed there – a whole breeding season wasted. Eventually a mocking bird might manage to break the shell, no easy task for such a slender bill though they persevered birdfully. On arid and impoverished Hood Island an albatross egg, no matter how rotten, was a real bonanza. So, too, was a sip of albatross oil which they stole whilst the parent was feeding its chick. The habitual oil thieves were as recognisable as motor mechanics used to be before garages became smart and clean and even more expensive.

Hood Island proved much tougher for us than Tower had been partly because of the cold and dismal mist which often blotted out the sun producing miserable grey weather. The north side seemed even colder and June, who evidently had begun to think like a booby, once remarked " I'm glad we nest on the south side".   On September 17th we opened our last tin of sausages and by then we would not even have entertained the thought of another year, if given the chance. Weak, maybe, for there was still plenty to do, but we had run out of steam. We had carried out 3,216 individual weighings of adult and chicks, countless measurements, examined hundreds of birds for moult, documented in detail the breeding behaviour and ecology of six species of seabirds, and filmed extensively. I doubt if we had taken a day off except during transit from Tower to Hood. Yet throughout it all, and it does seem remarkable, we had not ever quarrelled, sulked or even been fed up.

Sometimes, though, we were thoroughly fed up with the mocking birds. They

*Hood Island*
*northern coast*

were even more exasperating than the sea-lions. Our camp-followers on Hood were every bit as bold and innovative as those on Tower and the lucky band into whose territory we had so providentially blundered lost no time in investigating this wonderful windfall. Simply to survive, mockers have to be relentlessly opportunistic. One of them fell headfirst into a narrow jar half full of vinegar but by contorting into an extreme head-and-tail-up posture managed to avoid pickling itself. We dried it off in the oven and then down June's blouse. For days afterwards it ran everywhere because it was too stuck together to fly. Another in hot pursuit of a lizard, jumped into a bowl of washing-up water. Amusing though they were, the forty thieves severely frayed our tempers. They were forever putting out the wick flame of the paraffin fridge and they besieged our tent to such purpose that we dared not leave without barricading every possible entry hole. But it was a hopeless battle; they squirmed and wriggled to such good purpose, more like a mouse than a bird, that they always triumphed. Did an albatross come ashore to feed its chick? A mocker would be there to scrape up the slightest hint of spilt oil, assiduously working its bill sideways against the oil stained boulder. A molecule of oil for strenuous toil. Even sea-lion faeces and decomposing booby chicks were grist to the mill, but much more surprising, indeed innovative, was their attempt to eat live adult boobies. They stood underneath and pecked its cloaca until it bled and then delicately sipped the

*Lava lizard*

blood. On some Galapagos islands this repellent habit has been taken a step further by the small billed ground finches which attack the blood-filled quills of booby chicks, causing serious wounds in addition to exposing the victims to a nauseating plague of flies.

Mockers forage in groups. Their faces are individually recognisable by other mockers, though I must admit not by me, and they operate in a rigid hierarchy. One day several crowded around a tin scraping off bits of food, when an obviously dominant bird arrived and simply tossed the others aside like so much litter. In fact it used the same quick head-jerk that it employs to shift debris.

Usually, though, a new arrival ran up to another bird and if the latter's face was hidden it peered round from behind before attacking if dominant or running away if lower in the peck-order than its intended victim. Sometimes a subordinate bird made a mistake and pecked a dominant bird from behind, provoking swift and vigorous retribution. It would be delightful to make a detailed study of these fascinating birds and their relationships within the band. Group-territorial birds like mockers have especially interesting habits. Particularly when such

*Yellow warbler*

birds live in arid and demanding habitats, they may, like the Arabian babbler, breed co-operatively. So-called "helpers", probably related to the actual parents, lend a hand in feeding the brood. I would guess that mockers do so, too.

However interesting sea-lions and mocking birds, it was not they who had brought us to Hood Island but rather the blue-footed boobies and albatrosses. The boobies nested just behind our tent, as handy as the red-foot on Tower. There were no red-foots on Hood for they like warm blue water and Hood, in the south of the archipelago, is washed by the cold Humboldt current. This suits the blue-foot admirably for it prefers cold water areas such as those off parts of the Peruvian coast and islands in the Gulf of California. Ideally for us they were busily setting up nesting territories and courting. The blue-foot is the clown of the booby family. It flaunts its startlingly ultramarine webs and the male, in a virtuoso display, flies in and just before touchdown throws them up in front of himself so that they flash brilliantly against his white belly. They have heads and necks like chrysanthemum flowers, piercing yellow irises and a dark blue dagger-like bill. I disapprove of the school that writes "nothing could be more ridiculous than the courtship habits blah blah blah" but there is no denying that to our eyes the blue-foot does look droll. The male parades around flaunting his webs and the pair, facing each other, point their bills vertically upwards and literally turn their wings inside out so that the backs face the partner.

*Pair of blue-footed boobies*

*Blue-footed male advertising*

Once fairly engaged in nesting they may run into severe trouble, just like the other boobies that nest in these fickle waters. In a remarkable parallel to the happenings on Tower things changed dramatically on Hood in the course of a single month. The colony turned from a thriving nursery into a disaster area with dead chicks and abandoned eggs everywhere. At first it seems slightly odd to see parents idly resting and preening whilst their chicks are literally starving to death. Why, one may wonder, do they not spend all their time scouring the unproductive sea. But the answer is quite clear. Over their lifetime they will produce more offspring if they avoid serious stress even if this means letting a few chicks starve. Most chicks will die anyway before they become old enough to breed. Much better to ensure that the precious adults survive in good condition. The red-footed boobies and great frigatebirds on Tower Island had behaved in exactly the same way and for the same good reason.

*Blue-footed booby feeding its chick*

Some of the adults we caught had been ringed earlier by Miguel Castro and David Snow. David, meticulous as ever, had noted the contents of their nests and this enabled us to calculate that they were breeding once every nine months rather than once a year, as I had rather assumed they would. This stratagem was a response to the lack of seasons in the tropical Galapagos where, by and large, one time of year is as good, or more likely as bad, as any other. So the whole year is available. Breeding cycles of less-than-a-year or a year-and-a-bit are perfectly feasible. This is fine for the tropics but such non-seasonal strategies would have no chance of success on the Bass Rock, where they might result in birds trying to feed chicks in atrocious winter weather. But they are appropriate in the Galapagos. I found all this new and exciting.

It often happens that the implications of fieldwork which in itself may seem trivial, emerge only after later analysis. Sometimes, alas, the significance of the findings depends so heavily on statistical cosmetics that Joe Blogs may be handsomely forgiven for grave doubts. But, just occasionally, important bits of a jig-saw fall straight into place. In a very small way this happened to us when we saw a blue-footed booby's nest with two well-grown but unequally-sized chicks. Clearly the bigger chick had allowed its sibling to get enough food to live, but equally obviously it had collared most of the food for itself. At first this may seem much the same as the masked booby's practice of 'sibling murder' in which, soon after hatching, the bigger chick actively kills its sibling, but there is a world of difference. Because the older masked booby chick kills off the opposition straightaway it gets all the food from the word 'go', creating always a strictly one-chick brood. So at least it increases its chance of rearing something. On the other hand, the blue-foot can sometimes rear two chicks. The

*Siblicide in masked booby*

*Blue-footed booby with unequal size chicks*

masked booby has sibling-murder written into its genes; it does not depend on food being short at the time of brood-reduction. The blue-foot's method is far more flexible; if brood reduction does occur, it is by exclusion from food rather than direct murder. The younger sibling at least has a chance of survival whereas the masked booby junior is doomed if the first-born is still strong enough to kill it. Of course it may not be able to do so during food shortage and it is then that the second chick may survive.

Another piece of jig-saw fell into place when we noticed, with some surprise, how much smaller the male blue-foot was than his mate. But it fitted neatly with a significant sex-difference in foraging behaviour. Females seemed to fish further out at sea than males. Unfortunately we couldn't show this directly by going out there and looking, but once or twice we did see males seemingly commit suicide by diving headlong into the rocks. In fact they were plunging into very shallow rock-pools. This was just an indication that they could fish very close inshore in a way that the much heavier female could not. Though we couldn't check this directly we could at least time the duration of their fishing trips and in this way we found that the perky little male brought in food for the chicks more frequently than did the female. This did not mean that he worked harder, but it indicated that his routine differed from hers. So the chicks, which for the first couple of weeks needed frequent, albeit small feeds, benefited from the short trips of the male whilst later on the female, who is bigger, brought bigger feeds for the rapidly growing chicks. Alas for my 'Just so' story; another blue-footed booby buff (Hugh Drummond working in Mexico) found that his boobies did not behave like mine. Still, Mexico is not the Galapagos and there are such things as regional adaptations.

Trick dives were not its only accomplishment by a long chalk. We saw a colour-ringed male that a few hours earlier had been attending a colour-ringed chick, display to overflying females from a site more than fifty metres away. Ten minutes later he was back with the chick. Alas, our time was up before we could discover whether he began a new breeding attempt with a second female, leaving his old partner to finish rearing the chick on her own. This would be perfectly possible. The magnificent frigatebird in the Caribbean does precisely that but it would be a new discovery for a booby.

Early in October and for the last time, we happily began to pack a few things into our trusty wooden chests. Box number 4 was to contain all our precious notebooks, films and specimens – by far our greatest treasures and completely irreplaceable. But we had scarcely begun when the unmistakeable rig of old 'Beagle' appeared on the horizon.  Apart from an Ecuadorian lobster-fishing boat whose crew, after an apprais-ing look at June had asked us "porque no ninos" (why no children), we hadn't seen anybody for months. It turned out to be a surprise visit from Roger Perry, the Station's new Director, who, over a supper of (alas) tuna, our staple diet for weeks past, casu-ally mentioned that the Royal Yacht 'Britannia' was scheduled to call at Hood Island in a fortnight's time. By then, as it happened, we were down to half a tin of margarine, one of spam, some tuna and a little rice, flour and jam – not bad reckoning on June's part almost a year earlier and considering that we had not re-stocked in all that time. Paraffin was low, chairs broken and clothes in tatters, but the notebooks were full and the displays of three species of booby, great frigatebird, swallow-tailed gull and waved albatross had been captured on film – a good haul.

Punctually at 08.30 on November 4 1964 'Britannia' slid smoothly around Punta Suarez and made a perfect rendezvous with humble little 'Beagle II' hastening in under sail from the north. Good old Carl Angermeyer wasn't going to let the side down. Almost before 'Britannia's' propellers had stopped turning the rubber dinghies were on the move, bringing Prince Philip and a few others ashore. It seemed incon-gruous, in fact rather silly, to be standing on the beach of an uninhabited tropical is-land, looking like a dog's dinner and acting as the reception committee for a Royal visitor.

During his brief exploration of Punta Suarez HRH showed lively interest in the birds, especially the albatrosses which, although a bit reluctantly because of the time of day (they prefer early morning and late afternoon) put on a reasonable display, not exactly a Royal Command Performance but at least a face-saver. One of the shore-party (Corley-Smith, British Ambassador in Ecuador) later played an important part in helping the Galapagos Foundation. Another (Aubrey Buxton) eventually, as head of Anglia Television, sent Alan Root out to film Galapagos wildlife. The results were un-veiled at a Royal Command Performance in London that raised just about enough to fund a replacement for the beloved but ancient and leaky 'Beagle II'. So it was a fruit-ful visit. Unexpectedly, for it upset 'Britannia's' schedule and, no doubt, slightly miffed the assorted dignitories on board, Prince Philip invited us for lunch. We would have preferred another half-tin of tuna but a Royal invitation is not to be ignored. June, as ever, looked good, in HRH's words "as neat and tidy as the day she left civili-sation" but I suspect the Vice-Admirals, equerries and diplomats could have done without a bare-footed, bearded non-entity in patched shorts liberally splattered with

*HRH on Hood Island with JBN*

albatross vomit and blue-footed booby fertiliser. No matter; he obviously didn't care so neither did I. He was so charming and attentive to June sitting at his right-hand that she couldn't concentrate on the delicious meal. After a year on our stuff, that really did upset her. 'Britannia's' elegance took our breath away after the squalor of our final camp, with its sordid anti-sea-lion defences. Dove-grey carpets, softly gleaming woodwork, subdued lighting and panelled corridors hung with paintings of ships and yachts. A huge arrangement of irises decorated the dining table. Sheer luck had given us this bizarre but wonderful ending to our Galapagos year. For half-an-hour Cabin 12 was ours, with bathroom, lounge, writing table and easy chairs. But best of all, HRH agreed to take our films and notebooks back to London. This quite enormous favour eliminated my nightmare of losing the results of our year's work. A sizeable slice of our future rested in those soiled and battered notebooks and films, but we could hardly have carted everything with us to the Peruvian guano islands which were our next port of call. These legendary islands, source of immense wealth for a few but untold human suffering for others, were home to literally millions of boobies, cormorants and pelicans. We were jaded but virtually on their doorstep so it seemed unthinkable to miss them.

So, soon after 'Britannia' weighed anchor and disappeared round the corner we began to strike camp in readiness for the final return of 'Beagle'. Everything, down to the smallest scrap, had to be disposed of. A single bit of plastic, a tin can or a bottle would have been a desecration. Sea-lion droppings, yes, but Nelsons' litter, no. We found a deep crack in a nearby lava escarpment and there, at long last, we interred our faithful old paraffin stove and oven on which June had prepared so many feasts on the Bass Rock, then on Tower Island and here on Hood. Clearing up was a surprisingly long job but we set about it with gusto. It took a bit of juggling to have everything packed and ready to load onto 'Beagle' because it could have been several more days than had been planned. Snags on Santa Cruz could have delayed 'Beagle' and we had no means of communication. Meantime we had to eat and sleep, which meant using equipment. Yet it would have been out of the question to keep 'Beagle' hanging around whilst we finished packing. So we had to juggle everyday life with the readiness to leave more or less immediately.

Before quitting the Galapagos we wanted a quick trip back to Tower Island to see how the seabird breeding activities were going. As soon as we set foot on our familiar little beach who should fly down to meet us, head feathers raised in excitement, but our friendly little black finch. He went through his usual finger nipping routine by which he always asked for almonds and it seemed clear that after our five months absence he still recognised us. Alas, alas, - and I still feel bad about it- we had absolutely nothing with which to reward him. It was a miserable moment.

The frigatebirds had fared abysmally. Hardly any young had survived, partly because parents had left their small chicks unguarded whilst they were still vulnerable to the fierce Galapagos short-eared owl. But more bizarrely, some were killed by adult frigatebirds. Apparently, 'spare' adult males usurped the unguarded nests and attacked the helpless chicks. It was pathetic to see those vulnerable chicks, still small and downy, with their little cape of black feathers, inescapably tied to their whitened pads, all that remained of the miserable platform of twigs that frigatebirds call a nest. There they had to perch, exposed to the heat of the sun and fully open to whatever wanted to attack them. No wonder few survive. Add to this the fact that frigatebirds, over their adult lifetime, breed far less often than once every two years and it is hard, in the Galapagos at least, for them to maintain their numbers. Whether they are in fact doing so nobody knows. It would obviously have been better for the chick if one parent could have stayed on guard but the Galapagos is a harsh environment and both parents are needed as food-gatherers.

Although we never saw it happen, we strongly suspected the short-eared owls of attacking even well-grown booby chicks, badly lacerating the back of the head and neck, though I suppose it could have been the work of an intruding booby. Certainly, 'unemployed', that is, non-breeding masked boobies sometimes did wander around attacking unattended chicks but this automatically caused the chick to hide its bill beneath its body, which prevented further attack. The owls wreak havoc amongst the storm petrels on Tower and I once disturbed this fierce predator from a freshly killed whimbrel. It had eaten both eyes and the skin from the back of the head, which seemed an odd choice for starters.

The colony of Madeiran and Galapagos storm petrels on the horn of Darwin's Bay was still buzzing with activity just as it had been seven months earlier. There may have been successive waves of breeders, or perhaps a constant attendance of vast numbers of pre-breeders or non-breeding birds.

On Nov. 27th at 05.45 hrs. we finally left Tower Island. All the excitement of our landing the previous December had gone. It was simply a departure – no more than that. As an event, it was totally flat. We didn't even leave our bunks on 'Beagle' to watch its fuzzy outline recede into greyness. Yet, occasionally, I dream about that little beach and our crystal-clear mermaid's lagoon. Once out of the shelter of Darwin Bay the sea became quite rough and at one point water spurted through the planking into

Carl's bunk, bringing him on deck in double-quick time. Then it was Santa Cruz and the final preparation to fly from Baltra to Guayaquil. We left our wooden chests to find their way to Queen Elizabeth docks in London in their own good time, courtesy of a trusty Shipping Agent in Guayaquil. Our old wooden dinghy found an appreciative home with the equally aged Mr. Rambeck , a toothless, softly spoken and charming Norwegian; one of the early settlers. Doubtless its indestructible ironwood is still floating. Rolf Wittmer was the lucky one who got our faithful and, in the Galapagos of those days, invaluable paraffin refrigerator. Then it was Baltra and the endless hassle of negotiating a passage on the ancient DC10 flying to Guayaquil. It was packed with air-force personnel travelling to the mainland for Christmas and we were more than lucky to be allowed to join them. I fancy we had Carl to thank for that. Weight was critical if the old crate was to get airborne before it fell over the cliff at the end of the runway. Everybody was frantically weighing their dried fish, brightly varnished lobsters and, of all bloody things, crates full of lava and rocks! Goodness knows what they were for. Our own luggage was restricted to a tight 25kg so we went to enormous and quite unnecessary lengths to meet this, jettison-

*Tropicbird*

ing all sorts of things we would have liked to keep. Of course, nobody else did. Everybody piled on board carrying their brightly varnished lobsters which they carefully hung on the sagging electric cables which, totally unprotected, ran the length of the fuselage. The parachutes with which passengers were issued before boarding ran out just before the last man – me. Everybody piously crossed themselves. Maybe they knew something we didn't.  The pilot, who sported a fine Jimmy Edwards handlebar moustache, had landed in three giant hops and although he didn't take-off in like manner he only just managed to hoist the plane into the air before the runway disappeared at the edge of the cliff. The wrinkled sea crawled slowly beneath as we headed for the mainland – hard to miss. Passengers outward-bound from the mainland are not as lucky.

In Guayaquil we homed straight back to the old Pension Helbig where we were given a much better room than last time, almost a year ago. Did they really remember us? Outside, the stultifying heat was pierced by the incessant turmoil of horn-blaring traffic, car radios and all the babble and muted roar of a bustling South American city. Faintly, through it all, the weird call of the swallow-tail, the braying of the albatross, the yodelling of the frigatebird. But they were already fading. The turmoil of Guayaquil was now the reality. Still, it is not true that only the present is real. Maybe nobody ever said it was.

The present situation in the archipelago is dire. 'Galapagos at risk' by Graham Walkins and Felipe Cruz shows beyond question that the crux of the problem is the unsustainable increase in tourism and with it a huge surge in the resident population with its myriad demands on services.

We haven't been back to the Galapagos since 1964 and at that time the future looked bright. The Research Station had just opened, exciting conservation work and research lay ahead and eco-tourism was profitable but manageable. Now, however, the number of visitors has surged from 40,000 in 1990 (itself massively more than in the 1970s) to more than 140,000 in 2006 and has vastly increased since. This enormous growth without a long term strategy is heading towards massive markets and bigger and bigger boats with up to 500 passengers, dominated by multinational investors. Naturally enough the rewards from tourism have led to an influx of immigrants from Ecuador, which means more flights and cargo boats which are the biggest threat to native bio-diversity because of the alien species that they inevitably introduce, to say nothing of associated pollution. UNESCO now includes the Galapagos as a World Heritage Site in danger and the President of Ecuador has expressed great concern that the islands are at risk and has declared them a national priority for conservation.

As a consequence of the increased growth in population and the massive hike in the number of visitors there has been a surge in the number of introduced species of

plants and animals; 1321 in 2007, more than ten times the number registered in 1900. New vertebrate and invertebrate species arrive every year. The islands now have 748 species of introduced plants compared to 500 native species, and up to 60% of endemic plants are threatened. Meanwhile lobsters, sea cucumbers, groupers and sharks have plummeted as a direct result of crazily irresponsible exploitation. Sharks are being slaughtered by the tens of thousands just for their fins, which find a lucrative market in China. The Chinese are responsible, too, for the burgeoning trade in Galapagos sea-cucumbers which they consider to be an aphrodisiac. It is too crazy for words that China, of all countries, should be trying to boost its population. Sea cucumbers don't work, but they still pay the price. Ecuadorians from Guayaquil come out and hoover up the stocks and Park wardens try to stop them at their peril.

As if all this is not more than enough, there is the growing problem of investors creating markets for activities such as sport fishing, beach camping, kayaking, biking and even parachuting. As Walkins and Cruz point out, you can do those things in many other parts of the world. Why crash into the fragile Galapagos environment. The answer, of course, is profit. Walkins and Cruz put it well "analyses of market cycles in other tourism markets — identifies patterns of change that are self reinforcing and result in visitor reductions and lower revenues over the long term. Market cycling in tourism can eventually lead to complete collapse epitomising the history of Galapagos with the boom and bust of yet another lucrative product; with this collapse will come inevitable ecological degradation"

*Peruvian Seabirds*

## Chapter 4

# THE GUANO ISLANDS OF PERU

*"How can one paint a pen-picture of a thousand boobies diving, of a skyful of boobies which, in endless stream, poured downward into the sea. It was a curtain of darts, a barrage of birds."*

Frank Chapman 1933 'Autobiography of a Bird Lover'     New York

*"I have seen the penguin regiments of the far south, the courtship antics of the wandering albatross, a file of eighteen condors which passed me within a stone's toss and other marvellous sights in which birds held the centre of the stage, but nothing more exquisite than the pantomime of adult piqueros hanging on the wind above the cliffs of Guanape."*

Cushman Murphy 1936 'Oceanic Birds of South America' Macmillan New York

It was mid-morning on December 18th 1964. The twenty-five year-old 'Pacific Queen', a beamy old 60-footer, wallowed in an oily, turquoise-green swell outside the Peruvian port of Callao, on her way to the legendary guano islands. By the greatest of good-fortune- indeed a minor miracle - we were on board, the only passengers. To this day I honestly do not know why the illustrious Compania Administrado del Guano had paid the slightest attention to an obviously impecunious and unimportant biologist and his equally obscure wife. But attention, dignified and ineffably Hispano-American they had paid, to such effect that we were their guests, collected by taxi from our humble lodgings and now being transported in utmost comfort, with our own cabin, to the island of Guanape Norte where we were to be the guests of the hugely impoverished but impeccably courteous and generous Guardianes or keepers of this famous bird island.

The harbour was crowded with anchovy fishing boats exploiting the trillions of these small, sardine-like fish that swarm in the cold, rich waters of the Humboldt Current. The foothills of the Andes sloped mistily down to the calm Pacific Ocean, which

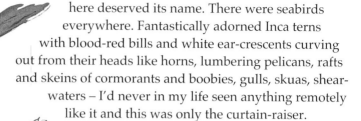

here deserved its name. There were seabirds everywhere. Fantastically adorned Inca terns with blood-red bills and white ear-crescents curving out from their heads like horns, lumbering pelicans, rafts and skeins of cormorants and boobies, gulls, skuas, shearwaters – I'd never in my life seen anything remotely like it and this was only the curtain-raiser.

The voyage took 30 blissful hours - more than a day cruising slowly down the most fabled seabird lane in the world. It was dreamlike, imagination made real. And at the end of it lay two islands, Guanape Norte and Guanape Sur, holding on their barren slopes many more seabirds than all the British seabird islands put together. On the walls of the deckhouse hung fresh meat and vegetables; on the deck two comfortable chairs and in the chairs two exultant travellers. In our cabin, clean bunks, on the table good, plain food. Outside, an endless procession of seabirds, ropes and skeins of them, thick, undulating lines, flocks, rafts, multitudes. What more could one desire? At that moment, nothing. It was completely perfect, better by a 100 country miles than a Linblad or any other organised cruise.

*Inca terns*

After Guanape Sur for three-minutes, we went on to Guanape Norte. These islands are utterly barren. The trite phrase "not a blade of grass or other vegetation" was literally true. There was guano, rock, birds and concrete - nothing else. The buildings, though substantial, were nearly empty. There were several bedrooms with no beds, a bathroom and a bare living room. All four guardianes lived in dirty, ramshackle huts; we had the guest quarters. Everything and everybody was smothered in

*The beautiful Inca tern*

*Part of the guanay cormorant colony on Guañape Norte*

fine guano dust. But the birds!  The island was a patchwork quilt with guanay cormorants, like pile on a carpet, forming discrete black masses and boobies, slightly less dense, lighter patches between them. Some areas which had recently been worked for guano were entirely deserted; bare ochre-coloured or creamy rock. In the air were ten, perhaps twenty thousand birds. Thousands more rafted on the sea and as dusk fell, unbroken streams and rivers poured back onto the rock in thick, endless ropes. In the tumult of a hundred thousand raucous voices all individual sounds were submerged. Rising and falling, the gabble remained as a constant background roar, as pervasive as the dust. Perhaps only the most populous of the great Northern seabird bazaars of Novaya Zemlya at their height, or the teeming penguin colonies of the Antarctic could hold a candle to this spectacle.

Next day we went with the Guardianes in an old rowing boat across the channel separating Guanape Norte from Sur. We embarked in enveloping mist, the nublita, and straightaway ran into a vast concourse of cormorants. They began to patter over the surface in their thousands before we could see them, a weird effect like muted applause from a huge, unseen audience. Soon they circled above the boat and we were in the centre of a moving cloud. Some Humboldt penguins added their voices to the din, like cows calling for a bull – most unbirdlike. South Guanape held immense numbers of Peruvian boobies; I reckoned more than a million and the Guardianes independently suggested the same rough figure. Imagine a million seabirds on a small island, little more than a rock. Many of the nests held two or three plump chicks cradled in their nests of grey concrete-like guano.

We rowed back to Guanape Norte in the beautiful evening light over a calm sea, the tops of the islands standing clear in the sun and tens of thousands of boobies hail-

*Part of the vast colony of piqueros on Guañape Norte*

storming down into the vast shoals of anchovies. Yet I knew that after a week or two at home all this beauty would seem infinitely remote and we would recall mainly the dirty food, the bugs and the smell. The truth is even sadder. Both the beauty and the squalor would diminish and the memory would fade into a miserable shadow of the event.

Christmas Eve saw us settled in our spartan quarters, June knitting by the feeble light of an ancient hurricane lamp whilst I scribbled a few notes. Our sacking-covered beds, festooned with cobwebs and guano, were infested with mites. Beneath the building the labyrinthine sea-caverns gurgled, slapped and soughed. Outside, the nublita enveloped the island in grey mist. When I was a boy, Christmas Eve saw the ginger wine steaming in a big earthenware yellow-lined bowl, set down by the kitchen range fire like Mr. Micawber's punch. The Christmas puddings would be standing in their cloth-covered basins and the Christmas fairy clinging to the topmost spire of the Christmas tree. I wouldn't have swapped places, though our present Christmas fare was soggy cabbage and lumpsuckers, third-class gristle and as glutinous as a slug. The Guardianes' kitchen, black with soot and cob-webbed, was filled with eye-watering smoke from the smouldering stove, probably from the dried stipes of seaweed which served as fuel, but their welcome was warm. Yet again I mourned our pathetic inability to communicate. If I had my time again, languages would be a priority. So Christmas Day dawned among the guano birds. I had often hoped that one day I would be here, amongst these teeming colonies. But the flesh dies hard in some of us and lumpsuckers fell just short of a plump goose or a leg of pork with crisp crackling. Yet we had merely to step outside and we were amongst seabirds by the million. The

*Part of the
vast colony of
piqueros on
Guañape Norte*

Guardiane reckoned there were 356,346 pairs of piqueros on Norte. I wouldn't vouch for the 46 or even the 6,346 but there were certainly hundreds of thousands. He classified them as 'dense' at 4 pairs per sq. m. which is pretty dense, but the cormorants considered this to be colossally extravagant. Almost a million packed themselves in at 7 pairs per sq. m.

There were no signs of starvation but when it does come, and it happens every few years, it devastates the entire population as Murphy graphically describes. The Nino phenomenon has been documented for nearly a century and was well known for hundreds of years before written records. Essentially it is a failure of the cold up-welling whose nutrients, brought up from the depths, sustain the food chain that culminates in the anchovies and their predators, very much including the boobies and cormorants. It cuts off food at source and for as long as the warm water persists there is nothing the seabirds can do except emigrate or starve. Most of them, including the

*Part of the
colony of
piqueros on
Guañape Norte*

emigrants, starve by the million. This phenomenon is not at all comparable to the sudden scarcities of fish which I described for Galapagos waters (though Ninos do affect them) because there the adult seabirds can usually survive by foraging more widely though most of the chicks may die and clutches of eggs are abandoned. Furthermore most Galapagos seabirds are not massively adapted to superabundant food as Peruvian seabirds are. Adult red-footed boobies in the Galapagos never die off massively, as Peruvian boobies do here.

At the time of our visit to Peru the teeming colonies were thriving, but twenty years later these marvellous seabird cities had been thoroughly devastated. The morbid effect of massive overfishing had been added to the natural catastrophes brought about by the Nino. It is hard to be dispassionate about this sort of thing. Despite all the signs and all the warnings from experts the senseless onslaught had continued. Astronomical numbers of anchovies had been taken, and when the seabirds were hit by a Nino event the depleted anchovy stocks were unable, afterwards, to fuel the usual rapid recovery, with the result that the guano birds which formerly numbered between 20 and 30 million, long remained at about a tenth of that figure. And of course the effect on the Peruvian fishing fleet was financially devastating.

*The bold piquero or Peruvian booby on Guañape Norte*

All this lay in the future. My first concern, booby behaviour, was a special thrill here because I had never seen a piquero. Its intriguing similarity to the blue-foot struck me at once even though they are distinct species. On the islands where both occur, they nest completely separately. Not just physically, but also in breeding ecology, they are so different but one could discern the dim outlines of their evolutionary past. One has to tread carefully here, for it is all too easy to construct 'Just so' stories that owe as much to the imagination as to the evidence. Yet much can be learned by constructing well-based scenarios. In the present case the first step is to look at the similarities and differences in their physical appearance, their ecology and their behaviour.

The piquero lacks the spiky chrysanthemum head and gaudy ultramarine webs of the blue-foot and whereas the male blue-foot is tiny, with an exceptionally long tail whilst the female is large and heavy, both male and female piquero are medium-sized without a noticeably long tail. Another important difference is that whereas the blue-foot breeds in several areas of the tropical Eastern Pacific the piquero is firmly restricted to the coasts and islands of Peru and northern Chile, right on the doorstep of the cold Humboldt current with its teeming millions of anchovies. In another, related, difference the piquero nests far more densely than the blue-foot and uses cliffs, which the latter rarely does. So far as behaviour goes, the piquero fights a lot more than the blue-foot does and there are many differences in their displays.

I could easily describe all this in detail and leave it at that as some people would prefer. But such an offering would be a bit like giving the reader a telephone directory – full of information, admittedly, but rather limited. We can at least do a bit better, by suggesting well-based interpretations.

The notable difference in size between male and female blue-foot suggests important differences in their fishing techniques and in the size of the fish they catch. A big and a little predator cannot take exactly the same size of prey. Just as the tiny cock sparrowhawk catches smaller birds than his large mate, so the male blue-foot couldn't handle the large fish which the female can. But aided by his lightness and long tail, which gives greater manoeuvrability, he can dive full tilt into the shallows in a way that would be suicidal for the female. Such differences, which open up more feeding niches, are obviously advantageous.

The piquero's case is very different. It feeds almost exclusively on anchovies; small fish which are normally unimaginably abundant. There would be little advantage to the booby having different-sized sexes since both are well suited for hunting anchovy.

Although vitally important, feeding behaviour is not the only thing that shapes the booby's physique; nesting requirements do so too. Here again the differences between the piquero and the blue-foot make good sense. The piquero is, or was, desperately crowded. Those barren rocks, glistening with icing sugar against the hazy blue sky and set amidst such rich fishing grounds comprise real-estate beyond compare. There is fierce competition not only within species but especially between cormorants, boobies and pelicans. These seabirds cluster thickly on the slopes, pack the broad ledges, cling to the least protruberance, enter the darkest gullies. The diving petrels even burrow beneath the surface crust – anywhere so long as they can somehow manage to rear their chicks. In the face of this competition the piquero has taken to ledges as well as slopes, in dense-packed masses. Even so, it often has to battle for its patch and overt fighting as well as vigorous threat is the order of the day. Both partners are involved and their equal size and aggression pays off just as in the gannet.

Things are very different for the blue-foot, dotted around in uncrowded colonies on flat ground or gentle slopes. With space rarely a problem, aggression is much rarer than in the piquero. There is less need for shared defence of territory and at least one Galapagos male held two territories simultaneously, though whether he had two mates we were not there long enough to find out. So the marked difference in size between the sexes doesn't reduce this booby's ability to defend its site.

Even some of the differences in courtship between piquero and blue-foot seem to make obvious sense, which is a lot more than you can say for all animal display! During early courtship the male blue-foot repeatedly circles the nesting colony, landing from time to time. Just before touch-down he throws his gaudy blue webs skywards in an acrobatic display. They flash, ultramarine blue, against his pristine white belly. If an albatross tried this it would be a disaster but for the agile little blue-foot it is no problem. One might easily suppose it was simply a braking manoeuvre prior to landing were it not so exaggerated and, critically, so tightly confined to the appropriate stage of the breeding cycle. It seemed such a nice bit of behaviour that I persuaded Colin Willock, of Anglia Television to include a bit about it in their Galapagos film in which the blue-foot featured. Getting a major TV company to delve into even that much ethology was a triumph and prompted Prince Philip, who was involved with the programme, to say " How does he know"?  The question was fair, but the answer is simple: observation and correlation. The piquero doesn't attempt this 'salute' as I called it. It would be suicidal when coming in to land on a crowded cliff-face.

*Male Peruvian booby, courtship display (sexual advertising). It resembles that of the closely related blue-foot but is less extreme*

Fitting the bits and pieces of an animal's behaviour and ecology together so that they make sense is fine, but as always some things don't fit in. For instance, I have no idea why the blue-foot has a chrysanthemum head and the piquero a white one, nor why the blue-foot has a yellow iris against the piquero's ruby red. But these attributes probably do have a function. Simply, more obvious things like the manoeuvrability imparted by a long tail make sense to us whilst other things may not. Natural selection is all-pervasive in shaping bodies and behaviour. Nothing just 'is'. It is so because –.

Describing the differences that now exist between piquero and blue-foot may or may not be intriguing but it doesn't tell us whether the ancestral booby which gave rise to the two present-day species was more like the present-day blue-foot or more like the piquero. They certainly DID share a common ancestor, but what was it like? My guess is that the piquero has diverged further from the ancestral sulid in becoming so highly adapted to the special conditions of the Humboldt and its environs.

Our privileged spell on these fantastic islands came to an abrupt end all too soon. On December 29th the 'Pacific Queen' arrived five days early and whisked us off to evil-smelling Chimbote whose fish-meal factories were busily and shamefully converting the lovely, nutritious anchovies into chicken food. Chicken food for obese North Americans whilst hundreds of thousands, if not millions, of Peruvians were starved of the very protein that the fish would have provided. To get these anchovies the surrounding sea was being over fished, exacerbating the natural disasters that periodically afflict millions of seabirds.

We got back to Lima just in time to see the city in its annual snowstorm, when at midday on New Year's Eve, the office staff tear up their old calendars and throw the pieces out of the window. The capricious fragments swirled and eddied like so many seabirds, closing the most pivotal and unrepeatable year of our lives.

*Alpine choughs*

Chapter 5

# BACK TO THE BASS

*"The Bass Rock's history is interwoven with that of its dominant feature, the gannet colony. This has always been the most accessible and best-known in Britain and the written record goes back to the Scotichronicon of 1447"*

SOC   1966   'Scottish Bird Islands'

After the Galapagos it felt good to be back on the old Bass. Had we ever been away? Had we really spent a year gazing out onto the sunny Pacific instead of this cold grey North Sea?  Little had changed, though change hung heavily in the air. Before we left in 1963 we had dismantled our hut and left the ruined chapel just as we had found it, except for the massive nettles, perhaps centuries old, which I had cleared in 1960 before laying the foundations for the hut. Or, to be more truthful, watching admiringly as George Robertson, the Principal Lighthouse Keeper, did so. We still had our Galapagos tent of stout tropical canvas and we tried it on the Bass, but that cold, windswept Rock is no place for a tent, so we gratefully accepted the Keepers' offer of the old forge. Its rear wall is actually part of the 13th century garrison. There is a furnace in one corner, still with the enormous leather bellows which in the old days the keepers had used for working metal. The furnace, which had consumed tons of coal, and the immaculate cast-iron kitchen range, had been chucked over the cliff when the living quarters were modernised. Sleeping in that unheated old forge with its corrugated tin roof in March or April was marginally less comfortable than sitting on a cliff ledge with the gannets. No wonder some of the poor old Covenanters who were incarcerated on the Bass in the late 1670s and 80s gave up the ghost. Rotten old Charles II. But to make up for the discomfort, fulmars cackled in the cavities of the old fortress wall whilst the sea sucked and surged below. In Estate Agent's jargon it had 'atmosphere'. June, though, was beginning to feel distinctly jaundiced about the Bass, but having shared so much it would have seemed strange to go there on my own. That sad day had to come, but not quite yet.

Many of our old gannet friends, recognised by their colour rings, were still there. One male had suffered an accident which left a strip of the horny outer layer of his upper mandible curling up like a rhino's horn.  With unfettered imagination we called him 'hornbill'. In time, his horn fell off, but the groove from which it had come darkened with pigment and this black line persisted for the rest of his long life. He never moved from his original site and became an old friend to be looked for each new season after his winter ocean wanderings. We last saw him when he was 37 years old. How many thousands of sea miles he must have covered, how many storms ridden out, how many thousands of fish caught. I wonder how he died. Maybe he became entangled in that fiendishly tough polypropylene netting, a horrible death which I have seen all too often. Maybe he was oiled, or finally overwhelmed by the load of toxins which gannets ingest from our polluted seas. Or maybe he died more naturally.

Our gear, stored in an old crate in the Keepers' paint store, was just as we had left it before we set off to the Galapagos, so we had pots and pans, spring balances and a 'snitching' pole with noose for catching gannets. It was all so familiar, so well-used, that now we were back on the dreich old Bass our idyllic lagoon on Tower Island, the warm sea and the hot sun faded – forgive the cliche - like a dream and we simply took up where we had left off. This re-rooting and re-routing was just what we needed to bring us back to earth.

Our gear may have been the same, but the Rock was changing fast. The gannets had spread inexorably up the north-west face below the little amphitheatre that must have been part of the original caldera. Here there had been a small colony of kittiwakes and one or two fulmars but in the late 1970s they mysteriously deserted. The astonishing increase of gannets had destroyed the waving thickets of the soft-stemmed grass 'Yorkshire Fog' on the north-west, and the turf and thin mantle of soil laid down in the previous hundred or more years. Over the millennia the underlying rock may have been exposed and covered again and again. These days the Bass resembles a gigantic, gannet-iced cake. How many centuries will pass before anybody again sees the Bass as we did in the sixties?

In bare statistics the Bass Rock is only about 350 feet high but so majestic that it seems hardly lower than its east coast rival, Ailsa Craig, three times as high. The Bass

*A sequence of pictures showing the spread of gannets on the Bass Rock*

A. 1961

B. 1999

C. 2005

D. 1984

*The 'wind shadow' by the foghorn was the last area to be occupied on the NW face*

85

E. 2000

*Explosion of numbers on the East, where formerly there were only a handful*

cliffs are highest and sheerest on the east, falling away on the south and rising again, round-shouldered, on the west. The gentle south face drops roughly in tiers from summit to sea. Below the cairn where once the Scottish lion proudly flew lies the ancient walled garden which used to grow food for the garrison. Then comes the ruined 14th Century chapel in which we lived for three magical years and below that the site of the garrison itself, now occupied by the lighthouse buildings and the remaining walls of the great 12th Century fortress. In 1902 many of the old garrison buildings were replaced by the lighthouse and its outbuildings. Many the nights I have slept with my head inches from the inner face of these historic garrison walls which, in season, now hold puffins and fulmars.

The final steep plunge to the rocky apron and the two landing places shows why the Bass was for centuries so impregnable. It would be a foolhardy invader who braved the cannon embrasures overlooking the inner landing and the way up from the outer one. Everywhere else is inaccessible except for the steep and narrow west gulley which at the bottom drops sheer into the sea. One man could defend it against a hundred. A few boulders rolled down would clear it in a trice, though there are few boulders on the Bass. History shows just how effectively this blend of natural and man-made obstacles defeated would-be intruders for when the fortress was dismantled in 1701 it had fallen only once in all those turbulent centuries. The old Rock must find modern life a bit tame!

Most famously in 1691 four Jacobites, imprisoned on the Bass, overpowered the sentry and joined by a dozen recruits held out against the Crown until 1694. Even then they emerged not only free but with army back pay of a few sovereigns. They would have made great trade unionists. No wonder the ancient historian, Hector Boece, described the Bass as "ane wonderful crag, risand within the sea, with so narrow and strait half (passage) that no schip nor boit may arrive but allanerlie at ane part of it. This crag is callet the Bass, unwinnabill by ingine of man". Shortly after the Jacobite debacle the garrison was dismantled and Sir Hew Dalrymple acquired the Rock for a purely nominal sum - I think a guinea. Renewed in the flesh over succeeding generations Sir Hew still owns it and with sons and grandsons, two "Hews"" among them, the Bass seems set fair to be the pride of a Hew Dalrymple for at least another century. More than one National body would love to get their hands on it but it is doing very well as it is.

At the base of the west gulley lies the entrance to a crooked passage maybe thirty or forty feet high, running from a small boulder beach and continuing beneath the rock to emerge on the east side as a high, vaulted cavern, in season alive with guillemots which whirr out in droves, tumbling pell-mell into the sea and fizzing down, pale blobs in the green depths. This cavern and passage is a dark, dank and gruesome place, a dreary recess "full of chill airs and dropping damp". In some years a grey seal or two hauls out there to drop a pup. When the Bass was beseiged by the English the sympathetic French always ready, as they still are, to do down the olde enemy, hid food and wine there for the Garrison. At one time the cave held the valuables of the Earl of Buccleugh and the records of the Church of Scotland. All I saw were one or two injured seabirds which had entered this gruesome place to die. What must it be like at midnight in a full-blown winter gale?

Just outside the cave entrance on the east side used to be a favourite nesting place for the sinuous, bottle green shag with its emerald eye, recurved crest and sepulchral croak. Dear old clownish shag with your huge feet, spare, wing-wrapped body and serpentine neck. They are so bold when they have chicks that you have to lift them off the nest and even then they quickly scramble back. Whilst we weighed the chicks the parents tugged at our clothes and pecked our feet, a feeble effort compared with the gannet's ferocious onslaught.

*Shag feeding a juvenile*

At the base of the cliffs on the west side a rocky platform juts out. Each year the morgue as we called it collects a grisly crop of unfortunate juvenile gannets that fall short of the sea. Thuggish great black-backs attend, ready to dispatch the casualties. They gather in late summer as ominous as Serengeti vultures. It seemed surprising that they were able to get through the tough skin beneath the feathers, even when the corpses were floating, but they managed it by tugging dementedly whilst back-peddling furiously.

If you live cheek by jowl with wildlife you become almost a part of it. This certainly comes through in my diaries in which I recorded all sorts of peripheral stuff simply to seal the memory. Early in April 1961 I climbed down the east cliffs to check for eggs on a broad ledge which is always the first place to receive its gannets and the last to lose them. To my chagrin I found 20 eggs, which meant I had certainly missed out on the earliest ones. Most were still fairly fresh, their white, limey shells streaked with dried blood, but a few were very dirty and could have been any age. The nests themselves often had to serve as footholds, as well as good hand-holds. Once or twice when I looked down between my legs I found myself nearer to the ultimate drop than I liked. I mention this to illustrate a point. Was it dangerous? Yes, up to a point. Should I have been forbidden to do it? Absolutely not.  Yet I'm pretty sure that today I would have to surmount all sorts of bureaucratic obstacles, if not put there by the University, then by some government department. We seem to have become dangerously risk-averse. Yet by using drugs and alcohol millions put their lives on the line without a qualm. Is there a connection or are they just ignorant?

Sometimes the fieldworker should ask himself whether what he is doing is justifiable, and it isn't always clear. As I climbed amongst the gannets to mark nests, eggs and chicks, all very necessary for me but inevitably causing disturbance, I often asked myself that question. When gannet chicks are tiny they snuggle on top of the adult's webs and a sudden movement by the parent can easily dislodge them. Even if they merely fall onto the nest or ledge below it is hellishly difficult to retrieve them without causing further mayhem. It can be a real headache. One chick which I dislodged landed in a nest below, much to the surprise of the adult. I slowly descended, retrieved it – not easy with irate gannets shouting and stabbing all around - and returned it to its parent. It was a lucky youngster. But if accidental doubling-up does occur and provided the chicks are about the same age, they will usually be reared as twins which grow just as well as singletons and lay down as much fat, which they need to live on during the first week or two of independence. This astonished David Lack of Oxford's Edward Grey Institute and prompted me to do a full-scale doubling-up experiment to prove the point, which wasn't as easy as it sounds. We had to weigh them throughout growth without disturbing the colony too much; a real headache. As the chicks grew bigger they became less and less inclined to put up with being handled, put into a bag and weighed. You couldn't blame them but it made life difficult. In the end we put up a small tent, divided it down the middle, made several quick forays into the colony, grabbed the unlucky youngsters and bundled them into one half of the tent. After weighing, we put them into the other half of the tent. Finally we returned them as quickly as we could to their nests and ran for it, leaving them to settle down in peace. It was traumatic work for us as well as the unlucky chicks but we

*The fulmar's simple 'cackle' display*

didn't lose a single one, which was a major triumph. When they reached 9 or 10 weeks of age we had to stop the weighing or they might easily have fledged prematurely, which would have been a disaster for them.

Although it was only February when I returned to the Rock in 1961 things were stirring. The air rang with the wild alarms of the herring gulls and thousands of gannets were already back on their sites. Fulmars were courting. For once it was a calm day. The sea sucked and slapped placidly against the inner landing and long brown straps of glistening kelp rose and fell sluggishly. The limpets and barnacles were fully exposed. Through the long centuries many dragging feet had landed at that precise spot. Sir Walter Stewart on his melancholy journey to the executioner's block at Stirling, reluctant soldiers manning this cold and lonely garrison, dejected Covenanters facing long incarceration in the lethally cold and damp dungeons and even modern lighthouse-keepers ill-pleased to leave hearth and home for the winter discomforts of the Bass. But my feet were eager enough! The following day was a beauty. All over Britain the temperature rose 10 or even 20 degrees above normal though the sea, still only 45

*Fulmar in flight*

degrees Fahrenheit, gripped like a vice and official survival time in it was only a few minutes. A fortnight earlier had been the hottest February day of the century at 65 degrees on the Air Ministry roof. I had been in Oxford preparing for the Bass and had thought of trying out my ancient 35mm. Exakta camera in the Park, but the presence of many amorous couples had scotched that idea. A year later to the very day, I was lying on top of the Bass trying to identify colour-ringed gannets on the N.W. slope below whilst squalls of sleet and snow filled my ears. This extreme variability of British weather is one good reason why our seabirds either lay their eggs over quite a long period to spread the risk of chicks hatching at a bad time, or else wait until quite late in the season when the weather is more favourable. If they get it wrong bad weather can easily disrupt the food supply or even kill their chicks outright.

There are plenty of boring or downright miserable moments during fieldwork but they rarely get a mention in the scientific accounts. Space in scientific journals is far too precious to waste on such trivia. The gannets just sat there, headless lumps sleeping snugly, totally indifferent to the weather. They sit on their cliff-ledges in an Arctic blizzard as comfortable as a cat on a cushion. If they can survive winter in the North Atlantic or the North Sea, a cold day on the Bass or on Bonaventure in Canada is nothing. On that bleak February day clouds of spray hurtled across the east landing. Tons of seawater foamed onto the unyielding basalt, separating into a myriad coursing rivulets to be whipped off the glistening rock by the blistering wind. The bulking East cliffs gleamed dully and the gannets, heavy as they are, were jostled and lifted bodily by the screaming wind. Sea temperature had fallen to 40 degrees yet only six days earlier it had been mild and sunny and flying insects had swarmed in the Garrison garden. A swallow could have eaten its fill in a few minutes. An early bee was out and about. Often, such warm winter weather is followed by mist and the next day the sea had been cocooned in cotton wool though the upper air was clear. The top of Berwick Law, seen from the Bass, loomed preposterously above the mist. Sleeping on the Rock in mid-winter in our unheated wooden hut with its thin, uninsulated walls was misery. We put layers of newspaper beneath our sleeping bags but the side next to the floorboards was always like a slab of frozen mutton. Worse, we had to sleep head to head.

During April and May we were busy from morning till night, what with behaviour checks, keeping track of egg-laying, catching, weighing and measuring adults and a score of other things. I was afraid that when I came to put the story together I would find a vital bit of the jig-saw missing so I did everything I could think of. Mike Cullen, my supervisor and second-in-command of the Behaviour Unit at Oxford, caustically dismissed my approach as "matchbox-top collecting" but it paid off, especially later on when I came to study the tropical relatives of the gannet. It was just one of the many ways in which Mike and I differed in our approach. His was the cerebral, clinical, selective and practical way; mine was woollier, more naturalistic, and so far as time and effort went more demanding. But it paid dividends.

During the nesting season gannets are amazingly bold. Indeed the males are violently aggressive and a hide may seem superfluous. But bold or not, a human presence makes them tense and anxious whereas for behaviour to be natural the bird has to be relaxed. That is why we always used a hide for our observations. Disturbance is

*Some gannet pair behaviour*

*Mutual preening reinforces the pair-bond*

*Copulation with nape-biting (a characteristic of gannets, absent in their relatives the boobies)*

a great problem in behavioural and ecological work and one which, often, is not given due weight. At first I didn't realise just how much observers affect the behaviour of the animals they study. Simply to enter a nesting colony, which is often necessary to gather basic information, can cause significant loss of eggs and chicks. A disturbed gannet may stand just a few inches off its egg or chick but enough for a herring gull to nip in and snatch it. Herring gulls are forever on the look-out for this, as I saw only too often. I once saw the loser in a fight between two kittiwakes haul itself onto the East Landing only to be attacked immediately by a herring gull, which would have killed it had I not intervened. All it took was that slight sign of weakness in the kitti-wake. Gulls are supreme scavengers too. Whenever we emptied edible rubbish over the cliffs they streamed in. On one occasion a gannet flew in with them and siezed

*The gannets meeting ceremony—*
*"mutual fencing"*

*Ritualised signal of impending*
*departure*

some chicken guts. It must have picked up the habit from following trawlers with gulls. A similar explanation probably applied to the gannet seen following a plough along with gulls.

Oddly enough, despite the rigours and restrictions of life on a Rock we were reluctant to go ashore. But in our first year (1961) I did attend the International Ethological Conference in Starnberg, Austria. This was my first ever trip abroad and it started badly. I shared a four-berth sleeper from Edinburgh to London with a chap who smoked black cigarettes incessantly and a fat, illiterate naval rating who snored like a pig. The cabin was like an oven which, combined with the stuffiness, put paid to any hope of sleep. The Eaton Hotel in London charged just over one pound for bed and breakfast. A seat at "The Crazy Gang" show cost 22p - a waste of money even at that. Next day the Dalmatian Express left Ostende at 4.50pm.and crawled through a dull and dreary Belgium, a fitting location, eventually, for the hub of the European Union. Thirteen tedious hours later, endured on a hard leather seat and no chance of a bite to eat until Munchen, we arrived. That evening two or three of us fell in with a charming German broadcaster, Herr Badestow. As a Bass Rock rustic I evidently fell under his spell - at any rate my diary enthuses about his views on national characteristics. Americans have a natural goodness, often doing the right thing for disinterested motives (this, remember, was 1961!) Germans and some other Europeans, especially the French, lack that impulse and the English are not really European. How very prescient.

On the traditional free day of the conference the erudite Professor Kalmus of London University took me under his wing and we went up into the Austrian Alps. I had imagined they would be a jagged version of the Scottish Cairngorms and similarly wild and empty. But at around 10,000 feet it was more like a Bank Holiday at Blackpool. The sky was too blue to be true and the Alpine choughs were delightful but the people! People sunbathing in bikinis, people in National costume, town's people in city clothes and hats sliding and slithering over the snowy scree. Somewhat underwhelmed we pushed on to the next ridge which plunged precipitously thousands of feet down to a lush valley. Picture-postcard pinewoods, log cabins with balconies and riotous window-boxes and, magically, no people. Maybe the others found it too far to walk or perhaps they just liked to be with others.

The conference ended with a party at Professor Usher's castle, beginning with a mock gosling being squashed in a large cheese press labelled: 'Stop! No more imprinting!' [newly-hatched goslings immediately follow or are imprinted on their parent]. In the courtyard a Bavarian band in National costume played behind a barricade of loaves, cheeses, German sausages and huge barrels of beer. The night slipped happily past until a midnight dip in the lake. Then bread, cheese and wine, and more dancing on the balconies to Professor Usher's accordion. At three in the morning it all drew to a natural close without any formalities, the perfect party. It was just a pity that June, who loves parties as much as I don't, missed it. Alas, those International Conferences, small enough to be intimate and awash with new ideas, are no more. I long ago stopped attending the large, complex and expensive modern versions.

Back on the Bass the season's crop of young gannets crowded the ledges and slopes protected from the gruesome massacres which they had endured for centuries.

*Fully grown juvenile, all down now pushed out by new feathers. Virtually a negative of adult plumage. Heavier than an adult*

*"Parliament goose" stage with wig*

In times past the Bass gannets provided a yearly harvest of fertiliser, feathers, eggs, fish and meat to the tune of a sizeable fortune in gold pieces. The men on the Rock eagerly awaited the gannets' return from their winter's feeding at sea, during which time they had been free from the exertion of flying back and forth between the colony and their feeding grounds, a round trip which can be up to 1000km. In February, when they return from their natural element the sea, to the perils of the land, they are desperately wary, prone to sudden 'panics' or 'dreads'. Suddenly the hubbub of strident voices dies away and the whole colony falls deathly silent before the gannets explode into mass flight as though the very devil is on their tails. Kittiwakes do the same thing. The odd feature is not that these oceanic birds should feel ill at ease on land but that the fear should strike all of them simultaneously. The sense of "Hell! What am I doing in this dangerous place?" is palpable. These early panic attacks have been known for

*By July the seasons' crop of young gannets crowd the slopes*

ages and the men of the Garrison used to take great pains not to disturb the newly arrived birds. Nearly 500 years ago Major wrote —" these geese in the spring of every
year return from the south to the rock of the Bass in flocks and for two or three days
during which the dwellers on the rock are careful to make no disturbing noise, the
birds fly around the rock".

Gannets find it difficult to land in a gale, pitching awkwardly onto their nests in a
flurry of dishevelled wings. That diabolical updraught against the cliffs tossed them
around like spindrift. The damage caused by awkward landings is the main cause of
natural death amongst breeding adults. Starvation, even in winter, seems rather uncommon, which is hardly surprising given that gannets can survive two or three
weeks with little or no food. Then they replenish their reserves by gargantuan feeding
when, eventually they do locate a shoal.

Once settled in at the beginning of the year, the gannets begin to repair their nests
which have subsided into sodden mounds or disintegrated during the storms of winter. Eventually comes the crop of eggs. The soldiers on the Bass must have known , as
the St. Kildans certainly did, that the gannet will replace a lost egg, sometimes twice,
which is not so surprising since the egg is small, only 3% of the female's weight, and
so not expensive to produce. In some seabirds the egg is many times as costly as that.
So taking the egg did not mean missing out on the chick, which would have been a
poor swap. The egg is highly palatable and a bit larger than a good-sized hen's. In the
old days they came from the cliffs by the basket-load, hundreds of them, their limy
shells still streaked with blood, a sure sign of freshness is soon lost once those muddy

webs have been wrapped round it. Albeit reluctantly, the gannets provided fish too, for when they panic they often regurgitate which makes it easier for them to take off. This is simply a gut reaction much as we sometimes feel sick with fear. Herring, sand-eel and mackerel emerged from their gullets "split-fresh". Although I have never seen a perfectly whole fish regurgitated, right down to the tip of the snout, (they are swallowed head-first so that the spines and fins lie flat against the body and consequently the head is the first bit to be digested) some were nearly whole and perfectly palatable. They were part-cooked, too. It was such fish, stolen from the honest gannet, that the avaricious Governor of the Bass sold dearly to the wretched Covenanters, 39 of whom were incarcerated on the Bass between 1672 and 1686.

Young gannets or "gugas" were — " esteemed a choice dish in Scotland and sold

*Gannets incubate their single egg beneath the webs*

very dear (9d. plucked). The beak is sharp pointede, the mouth very large and wide, the tongue very small, the eyes great, the foot hath four toes webbed together. It feeds upon mackerel and herring and the flesh of the young one smells and tastes strong of these fish. The other birds which nestle in the Bass are these: the scout (razorbill), the cattiwake (kittiwake), cormorant, scart (shag), and a bird called the turtledove, whole footed (webbed) and the foot red" (this may have been the tystie or black guillemot which nowadays doesn't nest on the Bass). Ray and Willoughby who wrote the above passage visited the Rock in August by which time the puffins and common guillemots, which they do not mention, may have left the Rock, for it is unlikely that there were none in those days.

Maybe something like a million gannets have been "harvested" (a cynical euphemism) from the Bass over the centuries. Man is a messy predator and his "harvest" included thousands more maimed and lost to drift away and die miserably. So there was undeniable justice in an accident which one year befell the "harvesters". It was their habit to club gannets and toss them down into the sea to be collected by boat. One such vessel, unnoticed by the business end of the operation, drifted close in beneath the cliffs. To the boatmens' consternation and terror they were suddenly assailed by a hailstorm of dead gannets which hurtled down, crashing into and around the boat. The surge of the sea and the wind drowned their frantic yells as the fruits of their comrades' grim labours made their final, mute protest. A fat guga can weigh 9 or 10 lbs. and the cliffs are high. The terrified men crouched in the rowing boat as the missiles rained about their ears. They were soon covered in gore   and feathers and presumably sick of gannets.

But what can compare with the following imaginary account from Stevenson's 'Catriona'. " At last the time came for Tam Dale to take young solans. This was a business he was well used with, he had been a cragsman frae a laddie and trustit nane but himself. So there was he, hingin by a line an speldering on the craig face whaur its hiest and steighest. Fower-twenty lads were on the tap, hauling the line and minding for his signals. But whaur Tam hung there was naething but the craig and the sea below and the Solans skirling and flying. It was a braw Spring morn and Tam whistled as he claught in the young geese. Mony's the time I've heard him tell of this experience and aye the swat ran upon the man. It chanced, ye see, that Tam keeked up, and he was awaur of a muckle solan pykin at the line. He thocht this by-ordinar and outside the creature's habits. He minded that ropes was unco' saft things and the Solan's neb and the Bass Rock unco' hard, and that twa hunner feet were raither mair than he would care to fa'. "Shoo", says Tam, "awa' bird!. Shoo, awa wi' ye." Says he. The Solan keekit' down into Tam's face and there was something unco' in the creature's ee. Just the ae keek it gied, and back to the rope. But now it wrocht and warstl't l'ke a thing dementit. There niver was the Solan that wrought as that Solan wrought, and it seemed to understand its emply brawly, birzihg' the saft rope between the neb of it and the crunckled jag o' stane.

There gaed a cauld stend o' fear into Tam's heart. "This thing is nae bird" thinks he. His een turnt backwards in his heid and the day gaed black aboot him. "If I get a dwam here", he thocht, "its by wi Tam Dale" and he signalled for the lads to pu' him up.

And it seemed the Solan understood the signals. For nae sooner was the signal made than he let be the rope, spied his wings, sqawked outloud and dashed draught at Tam Dale's 'een. Taqm had a knife, he gart the cauld steel glitter. And it seemed the Solan understood about knives, for nae sunner did the steel glint in the sun than he gied the ae squawk, but haigher, like a body disappointed and flegged off about the roundness of the craig, and Tam saw him nae mair. And as soon as that thing was gane Tam's heid drapt upon his shouther and they pulled him up like a tied corp, dadding on the Craig".

Late in the season the gannets become wary again just as they had been when they first returned to the Rock in February. Like their return, so their final departure, when they forsake the Rock which has tied them for so many months and return to their sea-wanderings, cannot be accurately predicted. The Rock may be practically deserted by the end of the first week in October only for thousands to flood back later, if only for a short time. Even early November can see a few adults circling the cliffs and a handful of forlorn black juveniles waiting for a last feed. And nowadays the Bass gannets are tending to lay two or three weeks later than they did twenty years ago. Correspond-

*Gannet in flight*

ingly, there are more youngsters still on the Rock in late October and November. But December, and in most years the best part of January sees the Rock completely clear of its faithful solans, which are scattered far and wide in the North Sea, North Atlantic, Baltic and Mediterranean.

During the many decades that I have been associated with the Bass gannets there have been many films and television programmes about them. They are so dramatic and telegenic that they cannot fail to enthral viewers. But I think the daddy of all programmes made by the BBC Natural History Unit was with Tony Soper as the narrator and me as the gannet. It was the height of summer but the Bass had put up its "No Visitors" sign. A cold and buffeting north-easterly had built up a nasty swell. A chilly drizzle reinforced the general gloom but this had to be ignored for ten million viewers were expecting a day's vicarious bird-watching live from the Bass. So Fred nosed out of North Berwick harbour in 'Sula II' with his cargo of cameramen and technicians and enough hot coffee to have filled the "Gannet Bath" (North Sea) of Beowolf. I had never seen anything like this 'Birdwatch' operation, involving a generator, miles of cable running clean over the top of the Rock, several video-cameras, space-age ranks of complex monitoring equipment, huge dishes to transmit the pictures to the control centre on the mainland and much else that baffled my Neolithic mind.

Tony and I sat at the edge of the colony and as the gannets minded their own business and fought, threatened, displayed and fed their chicks we talked about what they were doing and why. All that preparatory work by the technicians, all the thousands of feet of back-up film which the cameramen had taken in case a live broadcast turned out to be impossible, went into a mere half an hour's viewing. I dread to think what it all cost, but then think what the David Attenborough programmes cost and they are rightly deemed to be well worth it.

Towards the end of the 19th Century the number of gannets in Britain and indeed the world were at a historic low. The prevailing spirit was brutal: "evil days came upon the gannet when their principal value was thought to be only as a mark of sportsmen……. One year the whole west front of the Rock was depopulated". But as one threat recedes others materialise. Nowadays gannets suffer from oil, chemical pollution and from synthetic cordage from nets and lines. Each year thousands of youngsters on the Bass, and on other colonies, are growing up in nests dangerously fouled by this tough and unyielding material brought in as nest material by the adults. If only fishermen would refrain from throwing such stuff overboard.

In the old days gannet grease was gold. Hector Boece's theology may be forgotten but his description of gannet fat deserves better. "Within the bowellis of this geis is ane fatness of singulaire medicine, for it helis mony infirmeties, speciallie sik as cumis be gut (gout) and cater (catarrh) disceding (diseasing) in the haunches or lethes (groins) of men and women". Gannet grease certainly did my sore back a power of good, if only psychosomatically, and I'm sorry I have none left. The first dead guga I find is due for meltdown!

Bass gugas were no strangers to the Royal table for as Gurney relates, among the moorfowls, partridges, herons, cranes, dotterel, redshank and larks served on 4th September 1529, to James V at Edinburgh were "once Solares (Solan geese)". Nor were they strangers to the rough Irish labourers who came to the rich Lothian farms at harvest time and feasted on gugas roasted over huge fires at Canty Bay just across from the Bass. Roistering scenes.

In the late 1700s , after the Garrison had been de-fortified but long before the lighthouse was built, a few climbers lived for part of the year in a little hut on the Bass, probably situated where the lighthouse tower now stands. It seems that they sold liquor, bread and cheese to the visitors and to the vandals (they called themselves sportsmen) who came to the Rock for the pleasant pastime of shooting the nesting gannets. Those were indeed dark days for the solan geese, shot for "fun" as they tended their young which would then slowly starve to death whilst the vandals went home to their comfortable hearths.

In 1961 and 1962 Niko Tinbergen and I made a film about the gannet. We called it "Gannet City". Today the rankest amateur would despise our ancient cine camera but that beer-bottle-bottom lens filmed some excellent sequences which in purely behavioural terms have not been equalled. This may sound far-fetched but it boils down to how well the camera-man knows his subject. The difficult bit is getting an unbroken sequence from before the behaviour begins until after it has been completed. A professional cameraman, however good technically, cannot do that unless he has previously immersed himself in gannet behaviour and he will not have had time to do this. The obvious solution, which the BBC have used with David Attenborough and his team, is to combine an adequate fieldman with a professional camera crew. It took me eleven continuous hours tensed behind the camera lens waiting for the few seconds when the young gannet, on its maiden flight, made that irrevocable leap from the cliff on untried wings; a dramatic moment never filmed before.

Hans Kruuk's empathetic biography of Niko covers the period when Holland was occupied by Nazi Germany during World War II. Niko and some other prominent

Dutchmen were incarcerated in a concentration camp. Whilst he was with us on the Bass he recalled some of his experiences in his usual quiet, matter-of-fact tone. I remember one, in particular. It was about the Nazi reaction when the Dutch Resistance killed one of the German invaders. The reprisal was swift and savage. The dormitory door was kicked open at dead of night when morale was lowest, and jack-booted Germans stamped in. They stopped, quite arbitrarily, opposite one of the wooden bunks and the wretched occupant was given three hours to write farewell letters before being taken out and shot. It devastated the morale of the incarcerated Dutchmen, many of whom were highly distinguished men. It was useless terror for it did nothing to quell the Dutch resistance. Niko was a great raconteur with a fund of funny stories - no histrionics - just a well-modulated voice. One that tickled my

*A frieze of shags*

fancy was about the time when food, in particular meat, was scarce in Holland. At one meal his small daughter burst into tears. "Why, what's the matter"? All in one long breath and on a rising note: "Jack" (aged five) "told me that if I ate some of my meat he would tell me something funny. When I did he said "now I've got more meat than you"

After those three marvellous years living on the old Rock things could never feel the same again. I've been back scores of times, lived in a tent, bivvied in the old workshops and enjoyed the comfort of the renovated lighthouse quarters, but nothing gives the same sense of belonging as ten months at a stretch, taking all weather, tempestuous nights, magical dawns, migrants thronging the Rock, geese overhead, sickle moons and pitch black. I owe an awful lot to the Bass.

*Eiders courting*

Chapter 6

# BETWEEN ISLANDS

*'Take no thought for the morrow'*

We seemed to be taking this biblical injunction rather too seriously. By 1965 I was 33. We had spent three years on a rock in the North Sea immediately followed by a year on two uninhabited islands in the Galapagos. Our friends were well settled, mostly with families (and mortgages). All we had to show for our efforts was a pile – admittedly a big one – of well-travelled, guano spattered notebooks plastered with regurgitated fish scales from gannets, boobies and frigatebirds. I needed time; time to digest the contents and do something useful with them. Providentially the Leverhulme Foundation thought so too and on a memorably dreich day in a monstrously ugly block of offices in London, with Prof Otto Lowenstein in the Chair, they generously awarded me two whole year's sustenance. Just imagine! Two wonderful years to pick up the threads of the gannet work and dig into those hard-won Galapagos notebooks. No wonder I look back on the 1960s and the Leverhulme with great affection.

We were helped over the weeks immediately after our return by the enormous kindness of Niko and Lies Tinbergen in Oxford. They more or less treated us as family whilst we sorted ourselves out. The very first job was to retrieve our notebooks from Buckingham Palace where Prince Philip had taken them from Hood Island on 'Britannia'. This miraculous stroke of luck had averted the dire possibility that they would end up lost forever in some dark corner of a cavernous customs' shed whilst I beat my head against the brick wall of official incompetence and couldn't-care-lessness. There is such a thing as Sod's Law. Years later my film about Abbott's booby, shot with huge effort and some danger in the jungle tops of Christmas Island was 'lost' in Italian customs and never recovered, though perhaps it was not solely due to their perfectly honed incompetence. It may have had something to do with linguistics; 'The Booby Prize' conjuring up the wrong images. Anyway they lost it. It served me right for letting it out of my possession, but an International Conference had asked for it, so I sent it. Silly of me.

Perhaps I had a slightly closer relationship with Niko than his average D. Phil. student, partly because of the gannet filming we had done together on the Bass. He was passionately keen on bird photography, at which he excelled, and anybody who shared it was sure to strike a chord. At that time he was still in his fifties, spry, energetic and riding high, with an international reputation, plenty of money for research and a bunch of bright D. Phil. students, although when he first came to Oxford from Leiden it must have seemed a tricky move. His student and fellow Dutchman, Hans (Hyaena) Kruuk's fascinating biography of Niko ('Niko's Nature' 2003 OUP) tells how Leiden University had only recently appointed him to a Chair when he took it into his prematurely silver head to accept a lowly position as a mere Demonstrator at Oxford. To Leiden it must have seemed weirdly perverse, especially as Oxford did not exactly roll out the red carpet. Sir Alister Hardy, the Linacre Professor, was keen enough but J. W. Pringle, his austere successor, was demonstrably not a Niko fan.

Niko's first group of students included those two later luminaries Desmond Morris and Aubrey Manning. Hans Kruuk figured in the second batch along with Richard Dawkins who has now outshone the maestro in general acclaim, though not within the narrow discipline of Ethology. Oddly enough, very few of Niko's students turned into ethologists interested in the language and function of animal display. Most of them have leaned heavily towards ecological aspects of behaviour. For example, one can look at the way nests are spaced out in a gullery and ask what are the advantages of that particular spacing – how do the gulls benefit? That isn't behaviour; it is ecology. The behaviour is the fighting or display which brings that spacing about. Back in the mid-sixties, Gerard Baerends, Professor of Animal Behaviour at Leiden and one of Niko's earliest disciples, remarked that there were only about three of us at that time who were 'traditional' ethologists, describing and interpreting the language of behaviour. This approach has since become something of a dead end and much of my research probably wouldn't get funded or published today.

Unlike his famous Austrian friend and colleague, Konrad Lorenz, Niko seemed endearingly modest and, usually, gentle – not thrusting and overtly ambitious. And though his field studies seemed quite simple, he was outstandingly successful. But fashions change, in science as elsewhere, and possibly he would not now make such a deep mark. One thing is certain; he hated computers; though whether he would still hate them today is a different matter. Allied to his rare insight into the way animals work was a rare ability to communicate scientific findings simply. It was an object lesson to see him work his way through a student's imperfectly formulated research, teasing out the important points and suggesting further possibilities without putting anybody down.

One would imagine that the obvious value of photography – a good photo can be better than hundreds of words - would appeal to just about every ethologist or ecologist but it doesn't. David Lack, the Director of the Edward Grey Institute was passionate about ecology but I doubt if he ever took a photograph – they certainly don't illustrate his books. The same broadly applied to Vero Wynne-Edwards. Niko, though, was fanatical - he absolutely loved taking, developing and printing black and white photographs and his books testify to their excellence. He looked on it as a means of communicating and in the same vein he was a successful broadcaster and children's author. Although surprisingly long delayed, material success at last came his way but he was and always remained surprisingly frugal. Hans Kruuk puts this down to his

Calvinistic upbringing, but I suppose it could have been just an essentially simple nature. Although often full of fun he suffered deep depressions, presumably like those that destroyed his brother Luuk. It reminds one of the famous 'black dog' that plagued the great T.H. Huxley, Darwin's 'bulldog' and then his grandson Julian (brother of Aldous Huxley). Despite it all, Niko won through to share a Nobel Prize with Konrad Lorenz and Karl von Frisch. The latter famously interpreted the 'bee dance' by which honey bees tell their hive mates the direction and distance they must fly to find food-flowers. Niko, the small, modest Dutchman sat oddly alongside the ebullient, larger than life showman and one-time Nazi-supporter, Lorenz, who was interned by the Russians and according to his own account obtained valuable nutrients by eating insects, including cockroaches. After the war, as Hans movingly tells and to many people's surprise, Niko and Konrad resumed their deep friendship and collaboration which by common consent had marked the birth of the scientific study of animal behaviour in the 1930s. They were the founding fathers of ethology and the inspiration behind those early International Ethological conferences.

After one conference, held in Edinburgh in the nineties, Niko was dying to show Konrad the Bass Rock gannets. Of course he could have organised the trip himself but instead asked me and June to do it. Lorenz had never even seen a gannet let alone visited a gannetry in the full tumult of breeding. Luckily, that day we had the Rock and the gannets entirely to ourselves and we took him over to our observation colony on the northwest face. After our fill of gannet behaviour we lazed around and talked within the sheltering walls of the summit cairn where once the Scottish Lion proudly flew. At the time, the summit and most of the top of the Rock was clear of gannets, but

*On the summit of the Bass where the Scottish Lion used to fly when the rock was a fortress. Bryan, June, 'Gretl' Lorenz, Lies and Nico Timbergen, Konrad Lorenz*

below us the northwest slope where our hide used to be, was blanketed by thousands of snowy gannets, as always in a ferment of noisy activity. In a silly way I felt a sort of pride of ownership, as though I had brought them there and commanded a Royal Performance. Lorenz was mightily impressed and I suspect he even thought they topped his greylag geese.

That we so soon became solvent after our Galapagos venture we owed, as I said, entirely to the Leverhulme Foundation, whose grant enabled us to migrate from Oxford to Kingsbarns in Fife; not too far from the Bass and for me a home-coming. More than that, we now lived in style at Cambo House, the home of the Erskines, between St Andrews and the picturesque Fife village of Crail. In Spring millions of snowdrops carpet the woods and the long drive. Cambo, a massive pile, had a walled garden, tennis court, high ceilinged rooms and, in our flat, a canopied four-poster bed. A genuine old four-poster is about as far as you can get from a Woolworth's camp-bed. We fed the large open fire with driftwood carted from the nearby beach; yet again we were in clover.

Nor did it end there. Adrian Horridge, my mentor in St Andrews University Zoology Department was by now the Director of the Gatty Marine Lab and he generously allowed me to use its well-equipped darkroom for free – a huge bonus. Hundreds of wonderful images came to life as they developed in that tray of magic fluid – black and white photographs which, done commercially, would have cost me a fortune. So, amazingly, many of the dreams conjured up on those calm, velvety-black nights on tropical Tower as we sat chatting quietly under the tent verandah, soothed by the measured beat of the breaking waves, had magically materialised. Life was still a great adventure though we knew we would have to buckle down one day.

When I was a student at St Andrews, living in St Salvator's Hall of Residence, Dr Hank Taylor, the warden, used to advise his idle, hell-raising medical students (and some of them were diabolical) to 'nail your scrotums to the chair and do some work'. Good advice, in part, so I settled down with my field notebooks. It was a hard grind. Those scruffy notebooks stained with bird lime, albatross oil and fish slime and held together with string, embodied hundreds of hours of tedious toil and miles of ankle-cockling stumble across unforgiving lava. But it was all useless until sorted, interpreted and published. They held some fascinating accounts of feast and famine, sibling murder, courtship behaviour and evolution, but these stories wouldn't leap out unaided from the grubby pages. Then there were all those photographs and my cine-film, totally indispensable for illustrating complex behaviour. Niko used to say that scientists' assets are their research results. They may well be all there is to show for a lifetime's work. And they are unlikely to make you rich. But think of the lifestyle.

After a few months in venerable St Andrews and at the invitation of Prof Wynne-Edwards, we migrated to the eider-haunted Ythan and Culterty Field Station, part of Aberdeen's Zoology Department. Culterty, lively and informal under George Dunnet, buzzed with ideas. One of its brilliant schemes, though, was simply too ambitious. The aim was to chart the flow of energy, in hard figures (K/Cals) through the food chains operating in the estuary, starting with the capture of the sunlight's energy by green plants and then up the chain to the top predators such as fish-eating birds. The idea was similar to that of Oxford's Bureau of Animal Populations where the aim was to work out the energy flow in Whytham Woods, from detritus-eating worms through to owls and foxes via voles and shrews. Again, it was simply too complex with too many unquantifiables; a bit like economic forecasting. In any case, I was not involved – nor did I take any part in Culterty's long-term study of fulmars on the Orkney Island of Eynhallow. Fulmar ecology was already being well-covered by others and I was not free enough to undertake long-term detailed behavioural work. Fulmars are rather odd. They have only a very simple display 'cackle-and-headshake' which they use for just about everything, from staking a territory to courting. Some years later a Spanish student came to do an M Sc at Aberdeen and tried to make sense of fulmar display, but the fulmar won and she went home, little the wiser. It does seem odd that some seabirds with long lasting pair-bonds like gannets, evolve complex displays whilst others manage with extremely simple ones. It would be nice to have an explanation.

*Fulmar display*

As at many Universities there were cartloads of ecologists at Aberdeen but no ethologists. At Banchory in Deeside David Jenkins and Adam Watson were breaking exciting new ground with their fine study of red grouse. Old ideas about what controls the number of grouse on a heather moor were being turned upside down. Was it food (heather shoots) and if so how; or was it predators, shooting pressure; disease or what? Tradition, enthusiastically supported by gamekeepers of course, was that predators – 'vermin' – were mainly to blame for keeping grouse numbers lower than they could have been and ruthless war by gun, trap and poison was waged against raptors. David and Adam, though, showed something very different. Their findings were counter-intuitive and went dead against orthodox thinking. In a nutshell they found that the grouse themselves maintained the numbers which were appropriate for the amount of food available. But how did they do it? The answer turned out to be both elegant and simple. In autumn the dominant males excluded any surplus, socially inferior, birds from the moor. If it happened that a good breeding season had produced more grouse than the moor could adequately feed, some of them had to go. But how to decide who stayed and who went? It seems it was done by aggressive display rather than serious fighting. And the most unexpected and intriguing finding was that once the status of a bird had been decided as a winner or a loser, it stuck. There was no question of a socially inferior bird trying again and again to change its status. In fact, that apparent immutability of status, once accepted, seems widespread amongst social birds, from hens to jackdaws through geese. So the inferior grouse simply goes off to a poorer, more marginal part of the moor and, most likely, dies before spring. This obviously has implications for the size of the bag which it is deemed suitable to shoot. In many cases a higher number of grouse could quite sensibly have been shot

*Red grouse*

or indeed could have been taken by predators without harming the breeding stock. Most keepers don't believe it for one moment, because it is counter-intuitive. No matter that the research was detailed, long-term and had been critically evaluated. It is always hard to believe what you don't want to.

When I joined Aberdeen University in 1966 there was only one professor of Zoology and he occupied the prestigious, crown-appointed Regius Chair. Vero Copner Wynne-Edwards won fame in the cloistered world of academic zoology with his controversial book about the social mechanisms that control the size of animal populations, particularly their breeding rates. What prevents them from exploding and eating themselves out of house and home? Mankind has more or less exempted itself from natural regulatory mechanisms with the consequence that his numbers far exceed the optimum, with devastating results. It is a fact that animals mostly do avoid 'boom-and-bust' and Wynne-Edwards developed the idea that they achieved this by regulating their breeding output to suit the prevailing 'economic' situation. The grouse seemed a fine case in point, tailoring the number of territory holders to the amount of good feeding heather, and excluding the surplus before they caused a food shortage. It was perhaps the most rigorous study of its kind ever attempted.

Forget the idea, if you ever had it, that academics of whatever type, scientists or otherwise, are disinterested seekers after truth, dispassionate advocates subservient to the facts. Alas, no. At a meeting in Oxford at which Wynne-Edwards, invited by David Lack who held a very different opinion, put forward his theory on the regulation of animal numbers, an irate geneticist, Philip Shepherd, yelled from the back of the room 'Can't you see, you silly man?' Hardly the stuff of academic debate. But then, think of poor old Edward O. Wilson, the distinguished American academic, who had a bucket of water chucked over him by politically hostile students. All because he dared to suggest, quite correctly, that some human traits are genetically based. Wynne-Edwards' book was on the one hand hailed as the greatest work since Darwin's 'Origin of Species' and on the other dismissed as anecdotal and fatally flawed. Its weakness, which Wynne-Edwards admitted, was that it was difficult to understand how an individual could possibly benefit by withholding its own breeding effort simply to benefit others of the species who showed no such altruism. And such 'sacrifice' was basic to his theory. He believed he had resolved that problem in his second book.

As I said, he was the one and only professor in the Zoology Department whereas nowadays at Aberdeen and other universities 'Personal Chairs' have proliferated like ground elder. I have no idea why this has happened. Nor do I understand why so many civil servants, not necessarily with any teaching or University connection, are now graced with this academic title. They are scientific civil servants, not professors. It seems oddly inappropriate. Anyway, it was Prof Wynne-Edwards' interest in seabirds that led him to invite me to join his department.

When Britannia visited Hood Island in 1964 during our final days there, Prince Philip's equerries included an energetic naturalist, sportsman and TV executive, Aubrey Buxton, later Lord Buxton, former chairman of Anglia TV (he died, at 91, in 2009). Aubrey was extraordinary; a born leader with enormous drive who would have risen to the top in almost any profession, very much including politics. Apocryphal or not, the story is that when he was a Cambridge undergraduate he spoofed the university top brass into laying on a red-carpet reception for a visiting eastern potentate of great wealth. Who unmasked the stained and robed Aubrey I do not know, but the episode did little to further his university career, which came to a premature end.

Many years later he had a big hand in saving Christmas Island's Abbott's boobies from probable extinction by televising my film 'The Prize Booby' as one of Anglia's famous 'Survival' series. That film was shown also on Australian TV. Australia owns Christmas Island and, as it happened, Prince Philip visited Australia shortly afterwards and he mentioned the plight of Abbott's booby to the then Prime Minister, Malcolm Fraser. This together with the film played a big part in persuading the Australian Government to protect it. Wheels within wheels maybe, but in this instance it did a power of good.

Aubrey worked in London but with typical nonchalance he bought a wild, mountainous estate about as far away as he could get, on Sutherland's Atlantic shore beneath the massive bulk of Quinag, with its towering precipices. Amongst Ardvar's many delights were highland lochs with nesting red-throated divers and tiny offshore islets beloved of common gulls. At that time, Prince Philip was keen on bird photography and Ardvar was perfect for a few peaceful days watching and photographing

*HRH Prince Philip & Morton Boyd on Handa Island, 1972*

**109**

birds. We were invited along, mainly so that June could prepare breakfast before the cook arrived at this inaccessible spot. Morton Boyd and Neil Campbell of the Scottish Nature Conservancy joined us for a trip to Handa, a wonderful seabird island, now a reserve, lying a mere pebble's toss off the Sutherland coast. The cliffs teem with guillemots and the top of the island is patrolled by aggressive bonxies (great skuas) which showed no respect for the royal cranium. On the single track road to Handa a sad little morris minor was lying with two wheels in a ditch whilst two agitated, silver-haired matrons dithered nearby. Out jumped Prince charming and with a bit of help from us manhandled the car back on to the road; "try not to do that again".

The year passed pleasantly enough at Aberdeen before the next junction loomed. The signpost pointed two ways; both of them attractive. One said 'Research Fellowship, Aberdeen University'; the other 'Lectureship, Aberdeen University Zoology Department'

The first would take me to Christmas Island in the Indian Ocean, where an alluring and mysterious booby nested high in the jungle canopy. The second was the safe and sensible option 'a bird in the hand'. The booby won.

In 1967 Abbott's booby really was a mystery bird; a deep, commanding and most unbooby-like voice coming from the high green canopy of Christmas Island's jungle, its sole nesting place in the whole world. Hardly anything was known about it; such drawings as there were owed more to imagination than observation. Beautiful, rare Abbott's; inaccessible, highly aberrant (a large, web-footed seabird nesting in such a crazy place) and, although we didn't know it at the time, soon to be threatened with extinction; gone forever whilst still unknown, like so many other species this last century. Abbott's booby and Christmas Island were destined to dominate my life for years to come.

So we packed out bags and reset the compass for Christmas Island's coralline shores; the heaviest item in our baggage a 100' rope ladder.

*Curlew song flight*

*Andrew's frigates flying*

Chapter 7

# JUNGLE, SEABIRDS AND
# CHRISTMAS ISLAND

*"The face of parts of the sea cliff is sheer. In the remainder it has been undercut
at the base by the action of waves. Along the south coast, where the cliffs
receive the full force of the heavy south-east swells… it has been broken and cut
to leave a series of sharp-edged ridges and stack-like projections. Here….
numerous large caves like dental caries have been hollowed out. In many cases
the roof is split by fissures… so that when there is a strong sea running
through they throw up great columns of spray, sixty to eighty feet high"*

C.A. Gibson-Hill 1947 'Contributions to the Natural History of Christmas Island, in the Indian Ocean.
Bull. of Raffles Mus. 18

In 1967 'Orsova' of the great old shipping line 'Pacific and Orient' took us from
Southampton to within spitting distance of Christmas Island, the green jewel of the Indian Ocean and nothing like its atomic bomb namesake in the Pacific. It was deck
cricket on coconut matting, passengers versus crew; huge buffet lunches by the swimming pool as we slid deeper into the tropics, and independent trips ashore at Gibral-
tar, Bombay, Penang and Singapore. Those regular scheduled voyages, quite different
from today's glitzy holiday cruises, once reached every corner of the globe. Alas, they
are now defunct. Today, if you wanted to go round the world on scheduled sailings,
you would be hard pressed indeed. Some thirty years ago the author Gavin Young
tried it and even with trips on Chinese junks, fishing boats and naval vessels, to say
nothing of cargo boats and ordinary passenger vessels he barely managed it. And that
was with a cartload of important contacts wangling passages for him.

'Orsova' skirted Christmas Island close in but declined to put us ashore. It would
have meant disinterring our hold luggage, including that 30 metre rope ladder. We
planned to spend the best part of a year on the island and our quarry, the elusive Abbott's booby, nested high in the jungle trees. So we could only gaze at the unbroken

sweep of green canopy before carrying on to Fremantle where we picked up a phosphate cargo boat, a genuine rust bucket, from Bunbury, for the thousand mile journey back to Christmas Island.

Poor old Christmas Island. Reported by a homeward bound Merchantman of the East India Company on Christmas day 1643, the first known landing was by William Dampier in 1688. True to form, Britain claimed it but the island remained blessedly uninhabited for a further two hundred years until in 1888 a small group established a settlement in Flying Fish Cove, named after a British vessel whose landing party had first hacked their way to the top. It took them a fortnight. In some places deep, razor-fanged crevasses choked with formidably spiny creepers bar the way; it could take a day to cover a few yards.

*The fretted coastline and green jungle of Christmas Island.*

*Flying Fish Cove, Christmas Island. This was the area first settled in the late 1800s*

Before the settlement, two entrepreneurs (George Clunies Ross and John Murray) were given permission by the British government to find the phosphate. By this time, with the guano from the Peruvian seabird islands well past its hey-day phosphate was gold-dust. Six years later these canny fellows transferred their leases to the newly-created Christmas Island Phosphate Company and the exploitation began. It has never stopped. After many twists and turns, including a short period of occupation by the Japanese during the 1939-45 war, Britain eventually sold the island to Australia and New Zealand. The latter sold out to Australia and by the time the island loomed large on our horizon the Aussies had owned it since 1958 although the administration was still known as BPC – British Phosphate Commissioners - and BPC it will be throughout most of this account. For me, no letters resonate like BPC. They dominated a chunk of my life.

In 1967 virgin jungle covered huge tracts of the island, the worst terrain so hostile that it was 1972 before anybody managed to penetrate to the two westernmost corners. Imagine; a tiny island, known for hundreds of years and inhabited for three-quarters of a century and still partly untrodden. When the island came to light there were no indigenous inhabitants and probably never had been. No other island in the whole world, so heavily exploited as Christmas Island, could still boast such wonderful wildlife. Best of all, it was the sole nesting place of the enigmatic Abbott's booby, its voice a deep, bull-like bellow in the jungle top and its life still a mystery. What amazing luck simply to have the chance to put a large and spectacular seabird on the ornithological map in the second half of the 20th Century, and one that would complete our booby story. It nests high in the jungle canopy, widely scattered, the strangest booby in the world, though just how strange we didn't yet know!

*The first photograph ever taken of Abbott's booby at the nest. (1967)*

*Abbott's booby with 5 week-old chick*

Christmas Island is shaped a bit like a cartoon Scottie dog, or a bone with two big ends and a central shank. Its ramparts of needle sharp fretted limestone are breached here and there by small silver beaches, 'Lily', 'Dolly', 'Greta', 'Winifred' and 'Ethel', long-forgotten loves of long-forgotten employees of the BPC. The monsoonal Indian Ocean crashed against these hostile shores millions of years before man existed. In many places the sea deeply undercuts the cliffs, bulldozing blowholes from cavernous recesses and shooting plumes of spray high into the air with fearsome booming and hissing. Fishermen unlucky enough to fall into the sea in

*Female Abbott's booby (rosy pink bill) which had been grounded. Here on June's head*

such a place are doomed. Amongst the limestone pinnacles on the sea-terraces, brown boobies fly in through rainbow arcs of spray to nest in small pockets of earth. A bit further inland the glossy, salt-tolerant scaevola shrub merges into wickedly saw-toothed pandanus and then into the secondary growth which fringes the jungle proper, the dense green cloak of Christmas Island.

Today's sea-cliffs had fore-runners, long since lifted up by the buckling of sub-marine plates in the earth's crust. On one such segment of raised shore terrace, now deep in the jungle, stands a grove of mangroves which were once at sea-level. These ancient sea-cliffs, now inland, still attract boobies and frigatebirds which circle and soar in the updraughts. Brown boobies nest in sea-fretted cavities which are now on the edge of the inland cliff, the agile little males hanging adroitly in the updraughts, like hovering kestrels. We were particularly interested in the brown booby; indeed with Abbott's it completed our booby odyssey. The cliffs are furnished with so many knobs and 'jug-handles' that even tyros like us could climb them, though it was best not to look down. The forested shore terraces attract thousands of nesting red-footed boobies (the white form) and great frigatebirds.

*The Brown booby on Christmas island shore terrace. the chick is 4 weeks old*

The Christmas Island jungle can be idyllic. It can also be damp, dripping and desperately boring. On a calm, sunny morning, amongst the noble, high-soaring buttressed giants, the rich red earth alive with land-crabs, white-eyed warblers flitting through the foliage, ridiculously tame ground thrushes foraging in the leaf-litter, Imperial pigeons grunting and emerald doves cooing, fruit bats chittering and squealing and the air crackling with cicadas it is vibrantly alive. One could spend a lifetime studying a square metre of jungle or a single species of insect, crustacean, bat or bird. The riches of the jungle ecosystem would defeat a cartload of computers. Darwin and his great contemporary Wallace waxed ecstatic about jungles. But for me this particular jungle's crown jewel was the entire world's population of the booby named after its American collector, one W.L Abbott.

*Christmas Island jungle: widely buttressed for stability and jungle with rich secondary growth*

In 1967 we were fantastically lucky to find Christmas Island fairly close to pristine. For more than half a century since exploitation began it had survived reasonably intact. The riches which Murray and Clunies-Ross had detected at the end of the nineteenth century, a creamy powder lying snugly beneath the floor of the jungle, in deep pockets amongst the limestone pinnacles, had already been mined in the south of the island, leaving the rest unscathed. True, there had been a price to pay - the settlement itself in Flying Fish Cove, the narrow-gauge railway and the dirt road running from north to south of the island, the ugly drying plant high above the harbour, blanketing everything with dust, and the chute channelling the phosphate down into the cargo boats berthed on the sunlit Indian Ocean below. Thousands of tons of it shot straight into their holds in spectacular yellow dust storms. Yes, certainly a price, but to the stately Abbott's, cruising in high above the jungle from its fishing grounds far to the north-west, things looked pretty much as they did to its ancestors long before man set his calamitous foot on the island. Then, as now, it saw an undulating green canopy stretching from summit to surf-ringed shore, set against the wine-dark sea. In 1967 the settlement and mined-out areas were mere pin-pricks. Abbott's nesting areas in the centre and west of the island remained untouched, where ridges break the evenness of the jungle canopy and provide nicely emergent trees for landing and taking off. Touching down in a tree-top is dangerous for this long-winged, web-footed seabird. Jungles and boobies seems a crazy combination. If it misjudges and falls to the ground it simply sits there or falls prey to the massive pincers of the

*White eyed warbler*

*Fruit bats*

*Ground thrush*

118

robber crabs. It hasn't the faintest chance of fighting its way upwards through the canopy and it doesn't even try. A large seabird, accustomed to the vast, sunlit ocean, quietly starving on the floor of a jungle seems cruelly bizarre, but it happens. One may well wonder what possessed Abbott's to take to such a habitat in the first place. Could it have been competition for nesting space on the ground? We know so little about the world of 100,000 years ago.

So let the boisterous ex-patriates who never had it so good, or even dreamt of it, swill beer at the club. Let the Chinese play their endless games of majong amidst the cloying fumes of joss-sticks. The depths of the jungle were still Abbott's domain which nobody penetrated. The few Chinese and Malay fishermen who knew cunning routes to the coast, subtly marked by a blazed tree here and there, could barely see, let alone reach, the odd nest they might pass. Had Abbott's been as accessible as some of the brown and red-footed boobies

*Abbott's booby landing*

*Loading phosphate at Flying Fish Cove, Christmas Island*

the damnably predatory Malays would have killed every last one. Indeed they did kill all the red-footed and brown boobies in some parts of the island. Lily Beach was depressingly littered with booby heads amongst the debris of empty coke and Fanta cans. They also slaughtered the rare and wonderful Andrew's frigatebird and stuck their heads on twigs. Never mind that it nests nowhere else in the world. They probably didn't know and certainly wouldn't have cared. But Abbott's nested as it had always done, unheard and unmolested. Long ago, Abbott's used to nest thousands of miles away from Christmas Island, in the Pacific as well as the Indian Ocean, but dear old Homo sapiens killed them all. That it is now a relict species is due entirely to man.

*The endemic frigatebird of Christmas Island - Andrew's frigatebird, this one is a female with 2 month old chick*

All this was soon to change. Much of the beautiful virgin jungle was soon to feel the steel of the bulldozer and in a matter of days an entire ecosystem of wonderful complexity, the culmination of thousands of years of evolution, would be swept away, leaving behind a lunar desolation. The reason: calcium phosphate, fertiliser for the farmers of Australia, New Zealand and Malaysia. Phosphate mining on Christmas Island, funded by the Australian government helped farmers by selling the phosphate at 'cost', though cost included just about everything, from the generous salaries of a large work force, a vast amount of expensive equipment, shipping, housing, schools, hospital and post-office. Wisely, it even included funds for the eventual re-settlement in Australia of those Malays and Chinese who did not want to go home after years, or in some cases decades, on the island.

Five Commissioners drawn from various walks of life handled mining policy whilst a General Manager in Melbourne and an Island Manager implemented it. For many decades this had worked admirably in a benign, paternalistic fashion. Christmas Island was a tightly-knit little kingdom living in fine style on a lush, tropical island with elegance and luxury for the management and a standard of living for the workforce higher than they could ever have hoped for at home. So when we first knew the island in 1967 everybody felt well content. It had no airstrip, no television, no domestic telephone, no trade union and no trouble. Theft was virtually unknown with little need for the lock-up, which could hold two people. The Australian Government Representative, Charles Buffet, and a British ex-bobby, Mr. Farnsworth dispensed justice. Driving licences were signed with a thumbprint – I still have mine though the Scottish Police don't recognise it. The entire social atmosphere seemed kindly and good-natured. The Manager, though in no way obliged to help, generously gave us a free house and an ex-world war two jeep on its last legs, purely out of good-will.

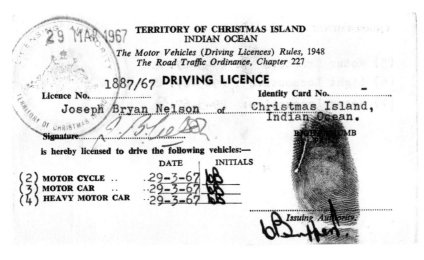

My driving licence for Christmas Island, 1967, Nos. 5, 6, 7 on reverse, qualify me to drive motor tractors, light locomotives and heavy locomotives

Alas, it was too good to last; the 'seventies were coming! Expansion hung menacingly in the air. Production, already high, seemed set to treble. Oddly enough it was the coincidence of our visit and the Company's detailed survey of the island in search of the remaining deposits that in the end saved Abbott's from extinction. Without the tracks which the surveyors bulldozed through the jungle to find where the phosphate was lurking, and which were fast closing-up, we would have had no hope of mapping the distribution of Abbott's nests. A hundred-odd square kilometres of jungle may not sound a lot but without those tracks we could never have searched the island. And without knowing where the nests were we could never have proved that, once again. Sod's Law triumphed, for just where Abbott's were thickest in the trees, Grade 'A' phosphate was lying snugly beneath the jungle floor. To extract it meant utterly destroying the habitat on which the world population of this fascinating seabird depended. That was the nub of the problem although it did not immediately seem so clear-cut. Before THAT revelation dawned we had to spend long hours tramping the grid of tracks and marking on a large-scale map exactly where we located a nest, heard a juvenile begging for food or a pair of adults greeting each other with their deep, resonant calls, more like a love-sick cow than a bird. Then, when we superimposed our nests onto a map of the phosphate deposits, which our surveyor friend David Powell had obtained, the problem became appallingly clear. But not until some five years later did the giant bulldozers actually began to rend and rip and push the trees over, bringing them crashing down with bewildered Abbott's boobies still sitting on their nests. Then the alarm bells really jangled. Exactly how many breeding adults were killed we will never know, but it must have been at least two hundred, which is a huge loss for such a tiny and vulnerable population with no chance of replenishment from elsewhere and no way of increasing its output of young.

*On Christmas Island, in 1967, the British Phosphate Commissioners kindly provided us with an old American jeep, invaluable for getting around in the jungle*

It was Dave who saved Abbott's. If he hadn't written to me in 1973, appalled by what he was seeing out there in the jungle, Abbott's booby would have been done for. David, a senior surveyor with the phosphate company, had become keenly interested in our work back in 1967. Before we left the island I asked him to keep an eye on some nests and tell me when the young boobies took their first flight. At that time nobody knew how long Abbott's breeding cycle took and we were not on the island long enough to find out. Most people would have been perfunctory about such a task, which involved a lot of time and hassle, but not David. As a long-time surveyor with BPC he knew the island better than any man alive and took endless trouble over Abbott's. He was as committed as me and sent regular updates. Some considerable time afterwards, when Anglia TV screened my film about Abbott's, Colin Willock, Anglia's brilliant script writer, always ready with a telling phrase, put it perfectly. As my full-screen shot of the head of Abbott's, with its large, dark, slightly protuberant and glowing eye slowly faded and its resonant voice lingered and died, Peter Scott drily narrated the final words —"will this rare and beautiful bird be wiped out for the sake of a handful of dust.". Some bird, some handful.

Don't blame BPC. They were on the island to mine phosphate and nobody had asked them to pay any attention to the wildlife. As one of their engineers – a fat and pallid German – put it, they mined phosphate, and didn't look after dicky birds. At that critical time, 1970-72, BPC had no Conservator to tell them where Abbott's nested, or even to point them to my published map which showed the distribution of nests. Naturally, after Dave's revelations, I had told BPC about my concerns but that meant very little. In all their 70 years of mining on the island they had been sovereign masters of their little kingdom, never compelled or even asked to pay any heed to environmental matters. So single mindedly did they pursue phosphate that they didn't even bother to salvage the rare and valuable hardwood trees which they felled by the thousand. Many hundreds of thousands of dollars-worth of excellent timber simply burned on huge funeral pyres. Beautiful, straight boles, 20, 30 or more metres long went up in smoke. Why? "Uneconomical to extract and ship". Uneconomical! By what mad criteria.? Time, machinery and transport neatly calculated against the presumed sale price of the timber. The usual crazy nonsense, even crazier in this case because they were not even supposed to be concerned with profit. So, given all this, why would they worry about Abbott's? Indeed, quite naturally but completely wrongly, they assumed as most people would have done, that the surviving boobies would simply move into the many intact areas of jungle. The fact that by sending bulldozers into Fields 17, 19, 20 etc. they were plundering the very heart of the entire world population of a critically rare and lovely seabird never entered the heads of the mining engineers. Even if it had, they would not have worried. Keeping up the flow of Grade 'A' phosphate was far more important. Had we not established the facts about Abbott's distribution in 1967, and had David and I not struck up a firm friendship, there might well have been no Christmas Island National Park today and little long-term hope for Abbott's. Everything flowed from our knowledge of where the boobies nested and how many there were.

But all that and many a tormenting twist and turn on the way was still to come. In 1973 David's agitated letters simply marked the beginning of a long and tortuous trail which, over more than a decade, took me back to the island several times, not easy from the North of Scotland. It led me, also, to the Commissioners' H.Q. in Melbourne and to the Australian Parliament in Canberra to give evidence about Abbott's booby to a Select Committee. And a year later, concern about Abbott's led to David's appointment by the Commissioners as their full-time Conservator on the island. They

*After felling (clearing) valuable hard wood trees are burned — a criminal waste!*

*A recently cleared field with newly exposed lime-stone and intact jungle in the background*

*A fairly recently mined area with stockpile*

124

Far left: *An ancient field, cleared many decades ago and still bare*

Left: *A field being rehabilitated by crushing limestone pinnacles and planting quick growing species to provide humus and shade for climax jungle trees*

could not have found anybody a tenth as good. Doggedly, he scoured the jungle, checking Abbott's numbers, distribution and, vitally important, their breeding success. "Not very good" was the answer. Almost certainly they do not begin to breed until they are several years old and even then they can't breed every year because it takes well over twelve months to rear their single chick. Add to this abysmally slow reproduction the fact that most of the young boobies die before they even leave the island and it becomes clear that Abbott's are not exactly rat-like breeders. But none of this was known until well into the 1970s

In 1974, at the express invitation of the Commissioners, I returned to the island with one simple aim – to establish whether Abbott's had reacted to the destruction of their nesting trees by moving into pristine but previously unoccupied jungle. If they had done so the pressure would be lifted from BPC because many of the areas which Abbott's ignored were of no interest to BPC. The Commissioners would happily leave them untouched. Yes, there would still have been the thorny issue of the nests in areas already earmarked for clearance, but this would have posed a lesser problem. Alas, the answer was all too clear. Abbott's was highly conservative and had not moved into pastures new. Conflict loomed.

Between our 1967 and '74 visits the island had changed dramatically for the worse. Just as in the Galapagos and countless other places, the sixties saw the end of an era and the beginning of frenetic global exploitation, of technological and communication advances, of habitat destruction, road building and pollution. And little did we dream of what lay ahead. On Christmas Island, alas, there now existed a full-sized landing strip for the new Charter flights from Singapore and Perth. Dear old 'Hoi-Houw' still sailed, though not for much longer, so for me it was back to her mahogany and polished brass. I boarded from Singapore's Clifford Pier as though only yesterday had June and I disembarked from our 1967 expedition. But if 'Hoi' seemed unchanged

Singapore did not. Old Chinatown had gone, replaced by Lee Huan U's towering office blocks, all chrome and glass, and by glossy department stores, testimony to his relentless modernisation.

I had just missed a berth on 'Navajo', a phosphate cargo boat, so with a few days to kill before 'Hoi' sailed I took a bus to Malacca, once British territory, to see the old fort. Rain poured down relentlessly and I felt aimless and ill-at-ease, as I always do in towns and on my own. A young man, friendly enough I suppose, tried to interest me in his sister. "Thanks, but I have a lovely wife" I said. "Where", he asked, peering around myopically. A simpering youth came up: "Do you like me?" "Like you" I snapped, "I don't even know you." Like the rain, all rather depressing. .

But I loved Singapore pavement food and Tiger beer. The tables stood on the road near to the cooking stalls. Large aluminium pots bubbled on cast-iron stoves; glowing charcoal kept the rice boiling nicely. In the cooking stalls huge shallow pans sat on fiercely hissing pressurised paraffin stoves with high tin wind-shields. On the shelves above this armoury, ranged a row of enamel pots of herbs, spices and nameless liquids. The artist in charge operated the set-up with masterly dexterity. Into the pans went some oil, then, rapidly, a ladleful of water. Up shot clouds of steam. On with the lid. Off again. Noodles, a splash from this bottle and from that, a shake from a herb-pot, a ladle-tip of spices and on with the lid again. After a moment it comes off and a fragrant cloud heralds the scooping out of a piping hot mixture. Onto a plate, garnish with this and that and within seconds it is on your table. Nothing could be fresher or more perfectly prepared.

What industry and toil went on in jam-packed Singapore and all with decency and cheerfulness. Crammed together in miniscule roof-top shanties they slaved for all the hours god sends - women breaking stones in the road in cruel midday heat, men pushing hand-carts piled sky-high with crates, bicycle rickshaws, sampongs heaped with rubber or sacks of this and that, toil, sweat and more toil. On Boat Quay the lighters unloaded sacks of meal, blocks of Malay rubber, charcoal, timber and miscellany. Whilst the coolies toiled a sleek Indonesian merchant with polished olive-wood face and black pill-box hat parked his gleaming Mercedes. Maybe re-incarnation will reverse the roles but what if we end up as woodlice or bacteria?

Sailing day was July 25th. 'Hoi-Houw' had been re-fitted since my previous voyage but the cabins were sweltering. As we headed out past Java mile after mile of forested hills reached down to the coast and seemed ideal for Abbott's. I eagerly scanned the canopy through binoculars, half expecting to see that familiar, rakish, long-winged form soaring above the trees: nothing.

By 5.30 in the morning we were lying off Christmas Island, which I had thought I would never see again, the warm tropical air full of birds – red-footed and brown boobies, frigatebirds, noddies and, high above, a lone Abbott's. He sailed out to sea, no part of the common herd. And there was David, his left arm, crushed during jungle felling, hanging as limp and useless as ever. At his side stood the Harbourmaster who hailed from, of all places, Dunbar, a stone's throw from the Bass Rock!

Next day David took me to some of the recently cleared areas in Abbott's heartland, the very places that had alerted him to the looming danger. What a dismal experience. Pinnacles of naked limestone reached to the sky, the deep hollows still cradling creamy phosphate, hundreds of tons that it had been judged uneconomical to recover. Those crazy economics again. Surely after devastating the jungle and killing so many Abbott's boobies they should at least have scraped the dish clean. In this depressing graveyard I could hear the rending and groaning of uprooted trees, the roar of the mammoth earth-movers, the clatter of mobile cranes with their iron buckets and the rumble of the huge Euclid trucks as they hauled the spoils away. And faintly through the uproar the resonant duetting of a pair of Abbott's re-united at their tree-top nest. Well, the trees are gone now.

After that depressing tour we scoured Tom's Ridge

*Dave Powell, conservation officer and surveyor, Christmas Island, hugely influential in saving Abbott's booby*

which - dismal prospect - was soon to be cleared. Back in 1967 it had been well-nigh impenetrable, the track burrowing through the greenery which dripped abominably after rain and shed prickly seeds down your neck. Soon it would be a desert. Even at that early stage in the devastation of the island, David was already planting seedlings of the climax jungle trees in some of the cleared areas, a thankless task, after the bulldozers had crushed and roughly levelled the naked limestone pinnacles. First he planted some quick-growing species for shade and, eventually, a miserable scattering of humus to give the indigenous jungle trees a start. They wouldn't have stood an earthly on the bare crushed limestone. He started a seedling tree-nursery by collecting growing seeds from the jungle. No professional could have tended his fragile charges more devotedly. His nursery would have done credit to any horticulturalist. Nobody else would have taken such trouble, but David loved the island. Others simply loved the money. To their credit, the Commissioners funded the materials and labour. Down at South Point he used washings from low-grade phosphate to back-fill huge crevasses between the pinnacles which, though getting on for a century old, were still naked. The dry slurry will be firm enough to plant secondary growth followed by tree seedlings. Maybe a hundred years from now South Point will again be forested thanks to David Powell. Well did he earn his MBE.

So, back to the familiar scents and sounds of the jungle where, in the dense trackless growth which covers much of the island, you could wander till you dropped. Even if you persevered and won through to the coast you would find yourself amongst thickets of pandanus, impassable spiky pinnacles and then the sea. And forget any idea of following the coast round until you hit the settlement. In many areas, certainly, you can walk freely between the soaring trees but in the worst places – some even on the doorstep of the settlement – deep gulleys, choked with dreadful thorny scrub, are virtually uncrossable. Where mining has opened up the jungle the native goshawks, much like our sparrowhawk, have become a real nuisance. One rose in front of me with its prey, a golden bosunbird chick.

*Backfilling an old mined-out area with slurry*

*Dave Powell's tree-nursery in which he reared climax jungle trees from seed, to replace jungle cleared for phosphate mining*

The ground was littered with its partly sheathed feathers and the remains of its last meal – a squid. The fierce little hawk must have dragged the bosunbird from its nest hole. It can kill small, unattended Abbott's chicks, too, even those of five or six weeks, though it cannot carry them away. Andrew's frigatebirds, too, also liked the clearings because the heated ground produces thermals on which they can soar and gain the height to swoop on incoming boobies to make them regurgitate their fish. If, as may happen, the frigatebird then forces an Abbott's booby down through the canopy and onto the jungle floor, it is doomed. One of the world's rarest seabirds killing another. Who could have predicted that particular ill-effect of jungle clearance? It is so easy to mess up ecosystems.

In 1974, nobody knew what happens to the young but unfledged booby during the monsoon period between November and March. David, surveying in the jungle, never saw an Abbott's and he simply thought that all the adults left the island and returned in spring. I knew this was impossible. When we left in September the chicks from eggs laid in May or June were only half grown. I had already shown that they would remain on the island until June or July of the year after that in which they had hatched. They could not survive unfed from November until the following March. Nor could they feed themselves. I knew all this for certain because when we first arrived on the island in February 1967 there were free-flying juveniles still being fed by their parents. Those juveniles simply must have hatched the previous year and overwintered on the island. They remained until June or July, visiting their nest-tree and were often fed. Then they abruptly disappeared – here today, gone for good the day after - as clear-cut as that. One thing that had made things difficult to unravel in 1967 was that, uniquely in the booby family and a fact unknown to us and anybody else, the juvenile Abbott's booby is indistinguishable from the adult male! Until we discovered this we were confused. Why should Abbott's be the only one with this unusual plumage trait? Of possible relevance, Abbott's, uniquely within the family, has adopted special behaviour to handle the aggression between mates. The aggressive jousting that occurs in other sulids would simply be too dangerous in the tree tops, where a slip could be fatal. Having eliminated this Abbott's has been free to dispense with distinctive juvenile plumage, whose function is to obviate aggression from adults, and move straight to adult colours, which presumably are adaptive.

*Female Abbott's booby orienting towards incoming mate*

*Pair greeting at nest, female has a pink bill, male a grey one*

David failed to see the boobies in winter because at that time the pairs with dependent offspring spend nearly all their time at sea, returning only briefly and infrequently. The juveniles, too, could easily be overlooked. They spend 90% of their time sitting quietly in the high canopy waiting for the magic moment when the parent returns with fish in its crop. Alas, for many of the youngsters waiting so stoically, that moment comes too late and they simply and quietly starve to death and fall to the jungle floor where land crabs quickly devour them.

Anybody familiar with the juvenile Abbott's begging call can easily track it down. It faces its newly arrived parent, always on the exact nest site even if all traces of the nest itself are long gone, bobbing its lowered head from side to side and shuffling its loosened wings whilst uttering a curiously harsh grating call. Even when the parent comes within touching range the youngster remains restrained, held back by an iron hand. What a contrast with the frenzied flailing onslaught of the juvenile red-footed booby which just about tears the head off its parent. Even when quite literally at death's door from starvation the young Abbott's uses its last dregs of strength to beg before falling off the branch, dead or dying. David has actually seen that happen. One imagines that at such a late stage a feed would not have saved it. By its curiously muted begging the young booby obeys an instinct forged over the millennia, for an all-out assault could easily knock the parent off the branch, a death warrant for both of them. That begging call is etched on my soul

After this visit and, as it turned out, for the last time, I boarded 'Hoi-Houw' for Singapore. As we pulled out of Flying Fish Cove in the late afternoon hordes of red-footed boobies drifted in. Many more rafted on the open sea and a flock of about a hundred were diving frenziedly. Brown boobies, frigatebirds and red-tailed tropicbirds were everywhere but there was not a single Abbott's. All these seabirds, quite apart from its endemic species, make the island unique. It supports more brown boobies than anywhere else in the Indian Ocean and, all told, is home to ten species of seabirds and getting on for a quarter-of-a-million individuals to say nothing of its endemic land-birds and its crabs. It is well worth International Heritage status.

My cabin soon began to stink, courtesy of a very dead 'mo-poke', a tiny owl found only on Christmas Island and named after its monotonous call. I wanted to keep it but all I had was a sheath knife and some common salt, and I am no Wallace, Darwin's great contemporary and, with him, the first to publish the theory of evolution by natural selection. Wallace collected hugely and was such an expert taxidermist that, like a friend I once had, he could skin specimens as he walked along. But it was not the malodorous owl that prised me from my bunk at dawn, rather the forlorn hope of encountering an Abbott's booby at sea and, amazingly, I did at 7 degrees 26'S and 105 degrees 10'E, about 200 km from Christmas Island. I saw two, the first-ever sighting at sea, although I'm sure others have seen them without identifying them. After all, before my 1967 photographs, existing drawings were hopelessly wrong. I already knew that each day the boobies returned to the island from the NW, so my sightings didn't tell me anything new, but it was still hugely exciting. I had become so accustomed to them in the jungle canopy that I had almost forgotten they were seabirds. Later I saw another off Java Head, about 400km from Christmas Island and in the region of a rich upwelling. It is not unusual for a seabird to forage hundreds of kilometres away from its breeding place but that sighting thrilled me.

I disembarked with some trepidation for I was on my way to the HQ of the BPC in Melbourne to screen my film of Abbott's booby which, at that time, had not been shown. It all stemmed from that desperate letter of Dave Powell's the year before,

*Abbott's boobies wing waving gesture*

describing the carnage amongst Abbott's during jungle-clearing. There was I, back in Scotland, moaning about the unfolding tragedy but doing nothing. Yet I had hundreds of metres of unique film – the first ever footage of Abbott's. I had taken it whilst balancing high in a jungle tree, with my heavy wooden Arriflex tripod and my hand-cranked Bolex cine-camera, left over from our Galapagos expedition. Abbott's at the nest, greeting its mate with that wonderful, ecstatic display, Abbott's feeding its chick with huge fish, so heavy that the youngster couldn't cope and had to drop it, and Abbott's building long, leafy sprays into the nest. And I had filmed the devastating destruction of the jungle, the remorseless bulldozers uprooting noble trees, buttressed Planchonella, straight-boled Eugenia, with their towering canopies and massive tree-ferns. So the hours of lugging a heavy old camera, large lens and wooden tripod through the jungle, up a cliff and then up a tree, bore fruit. The vigils had been long, waiting for that magic moment when the parent from far out in the Indian Ocean winged its way home and the re-united pair set the jungle echoing with their ecstatic dueting. And there were shots of Abbott's just sitting peacefully, gazing across the jungle canopy with those large, glowing and slightly protuberant eyes.

The Commissioners already knew a little about Abbott's but to see it more or less in the flesh, beautiful and dignified, and to actually watch the destruction of its wonderful jungle habitat was a world away from reading a dry annual report. They were genuinely concerned by what they saw but at that early stage none of us realised just how far-reaching the conflict between mining and conservation would become. They wanted to help. They laid down mining policy though answerable to the government not to profit-minded shareholders. So the way forward was through the government.

ANPWS made all the running through their tough and wholly admirable second-in-command, Neville Gare who, though strictly unofficially, kept me abreast of various governmental machinations. So one day I found myself in Parliament House, Canberra, giving evidence to the Standing Committee on Science and Technology. I had simply to ensure that they fully understood Abbott's booby! How many there were, where on the island they nested, how many years it took for a pair simply to rear enough young to replace themselves. Crucially, I had to convince them that it was confined to the very areas that the miners wanted to clear-fell. Looking back, this started the process which culminated in the proclamation of the Christmas Island National Park and saved much of the jungle and Abbott's booby.

So I returned to the raw chill of an approaching Aberdeenshire winter feeling a lot more optimistic. Things were moving in the right direction. I had no inkling that in a couple more years I would be back on the island, and by no means for the last time.

In 1976 the Commissioners invited me to see how their new policy of 'selective clearing' was working. Their idea was to ring-fence the trees with nests in them whilst at the same time felling the surrounding trees in the phosphate-rich areas so that mining could proceed. This produced some truly bizarre sights. In the midst of general devastation, mangled roots, shattered limestone pinnacles, churned up humus, there might stand a single surviving tree, gaunt, scarred and inevitably doomed, with an Abbott's booby nest high in the branches. Amazingly, the adults had persisted during the shattering upheaval, the bull-dozers, earth-movers and the falling trees and continued to sit on their egg. What a contrast to the serene tranquillity of the jungle canopy rolling unbrokenly into the far green distance. One can only marvel at the faithful birds, apparently indifferent to the chaos. Some of those doomed trees, jealously guarded by David, had been sacrosanct for so long that all the surrounding phosphate had been mined, smothering the tree with white dust. This 'selective clearing' undoubtedly saved many boobies but it did not save their all-important habitat and so could not save Abbott's. The fact is that there remains just this one tiny population of this seabird in the world and many individuals have already been lost. Abbott's breeds desperately slowly. It isn't like a rat, a rabbit or an Imperial pigeon. It cannot lay more than its normal single egg per clutch, nor can it breed more frequently. If you kill breeding adults by bulldozing them and their nesting trees you simply get to a point where so few remain that extinction becomes inevitable because the population is operating from an impossibly tiny base. Phosphate mining in Abbott's breeding areas was exerting precisely that lethal attrition.

So Christmas Island's future looked dire. Politicians were making silly proposals for 'using' the island after mining had been stopped. By this time BPC was picking up astronomical bills for Government administration, almost all of which it did not want and had never needed in those halcyon days – how distant they now seemed – when we had first visited the island and the Commissioners had been Lords of the Universe. The Commission itself ran everything. But once politicians get their acquisitive little hands on anything look out for growth, waste, bureaucracy and empire building. Once Canberra got its teeth into the island everything changed. In a single year the Australian government presented the Company with a bill for nearly three million dollars. This put enormous pressure on the jungle. More and more phosphate had to be mined, and quickly. Here David was uniquely valuable. For instance, there were some excellent tracts of jungle on a ridge above Field 22 but the miners wanted to clear a huge swathe of it in order to drive a road through. By linking up existing cleared areas David propose a route that sacrificed no jungle of value - the perfect solution but the mining department would simply have ignored anybody else.

BPC sometimes muddled through, making mistakes in assessing the potential of an area and the time it would take to mine it. They worked hand-to-mouth, quite forgetting that they had already stockpiled thousands of tons. They allowed stockpiles to become overgrown and kept inadequate records. People moved on and only David knew their whereabouts. Occasionally they overlooked the potential of a partly-mined area in favour of opening up a new one, putting yet further pressure on uncleared jungle.

By 1977 BPC was determined to sort out the conservation issues surrounding mining and Abbott's once and for all, so they generously agreed to fund, jointly with ANPWS, an expensive long-term study. From my original work, buttressed by David's observations, it was clear that to get a reasonable idea of how often and how successfully Abbott's booby breeds we needed to look at a minimum of three complete breeding cycles, which usually take two years each. So a three-man, six-year study was set up. It was led by a former Ph.D student of mine, Barry Reville who had done pioneering work on the frigatebirds of Aldabra in the Seychelles. These birds nest in the mangrove swamps, a hellish habitat to work in. Barry had been resourceful, meticulous and good at analysing data, just the qualities needed for work on Abbotts'.

As a bonus, he was a home-grown Aussie rather than a 'pommie bastard'. Barry and his co-workers, Geoff Tranter and Hugh Yorkston made an excellent team. This new study was a far cry from our grossly under-funded pioneering work of 1967, all sweat, string and sealing wax. I had scavenged on BPC's rubbish dump for bits of packing case to make rungs for scaling the trees but this de-luxe version had an office, computer and all the equipment it needed including new, four-track vehicles. It was the Bass Rock and Galapagos syndrome all over again! We blazed the trail at minimum cost but with maximum effect and others followed, funded up to their ears. Of course it was my own choice to do things my way and I don't regret it, but I can still moan!

A panel of three was set up to oversee the project. Mike Cullen, now Professor of Zoology at Monash University, Derek Ovington, Director of ANPWS and me. Our job was to supervise and advise and we met every year on the island – a lot easier for the others than for me.

We still faced some questions about Abbott's. Most of all we needed to verify my belief that it really was restricted to certain parts of the island. What about all those wonderful stretches of jungle that never saw an Abbott's? Could we have been wrong? The final answer, though, was plain. A displaced bird simply moved to the margin of the clearing – in other words they remained as close as possible to where they had been before calamity struck. They ignored the vast stretches of pristine jungle in which we had never seen Abbott's. We found this pattern again and again even though the margins of the clearings were now vastly degraded habitat in which their nesting attempts often failed.

Although this answered one simple question it raised others. Boobies fared badly at the fringes of clearings, but how far back into the virgin jungle did this 'edge effect' go? Did the partners of a displaced pair manage to re-unite at a new site or did they split up? If they did split, the implications for productivity were serious, for they normally pair for life and if they had to forge new pair-bonds which takes time, it would mean that they would produce fewer chicks in their lifetime. Things move slowly in Abbott's world. How many potential breeding years would they lose? Were their new sites as successful as their old ones had been? Questions, questions and more questions and all of them hugely difficult and time-consuming to answer, especially because Abbott's are so inaccessible and cannot readily be caught and marked so that you can identify individuals – dead easy in gannets, hellishly difficult in Abbott's. David had already discovered that in some years almost all the juveniles die in cyclones or by starvation. We had to find out the frequency of these bad years and the productivity of the good ones; more than enough challenges for the new team.

By May 1984 the island had become a political hot potato in Australia. In 1980 a Royal Commission had tried to be fair both to mining interests and to Conservation – difficult and inevitably curtailing BPC activities. Compounding the problem in a big way was the relatively new and confrontational Union of Christmas Island workers. Trouble was brewing in the form of a militant Pom who had arrived on the island and, quickly sensing an opportunity, had made himself powerful, stirring up unrest. The Government's perceived tendency to support the Union even when it was clearly at fault exasperated BPC (now called 'Phosphate Mining, Christmas Island') and led to the resignation of the Island Manager. Amidst the turmoil PMCI announced that it would start operations on the shore terraces, which had always been sacrosanct and were in any case fiendishly difficult terrain. The terraces supported huge concentrations of frigatebirds and red-footed boobies as well as golden bosunbirds, a rare and endemic form of tropicbird. Happily this project never got off the ground.

At least in Conservation circles, Abbott's booby had by now moved from embattled obscurity to international prominence. The IUCN had issued a magnificent set of commemorative stamps using my photographs, which were still the only ones available. Anglia TV had at last screened my film in the UK and Australia. Most of the island had become a designated National Park, thanks to ANPWS, and much of its wonderful jungle still remained. Abbott's booby, virtually unknown when we went to the island in 1967, was now famous in Australia. Even the then PM, Malcolm Fraser, had asked to be kept personally informed about its status. Eventually Hugh Yorkston drafted a proposal to make the island a World Heritage Site, which it richly deserves.

By 1984 Barrie Reville's team was well into its stride. An intriguing part of the study involved continuous surveillance of three nests using cine cameras. By exposing a single frame at standard intervals a roll of film lasted for 24 hours or more. The film then had to be retrieved and replaced high in the tree, which meant using a clever system of ropes and pulleys to lower and then replace the camera but the main snag was the enormous amount of time needed to analyse thousands of feet of film frame by frame. Most of it still remains unexamined and likely to stay so even if anybody can still find it. Only someone intimately acquainted with Abbott's behaviour could make sense of it and so far as I know I am the only one with the time and knowledge. That is the snag with that sort of data. It seems a good idea at the time and I did something similar with Cape Kidnapper's gannets but I have never properly analysed it. Other things tend to take priority and once it gets on the back burner it tends to stay there.

Back In 1967 one of our earliest observations had been that, every day, the free-flying juveniles returned with clockwork regularity to a precise place in a particular tree – a crotch or a fork in a branch – where, eventually, it was fed by an adult. It was always fed there and never in the canopy. That simple observation had puzzled us for the simple reason that the juvenile is an almost exact replica of the adult male. The books didn't know that and nor did I. No other juvenile sulid looks like its father! So it looked as though one male was being fed by another and it took us some time to work out what was going on. It hardly helped that, at the time, it was early spring when by all the normal rules there shouldn't have been any juveniles remaining on the island. Only later did we discover that, as I said earlier, the Abbott's booby chick hatches around July of one year and is fed in the nesting tree (and only there) until the following July, at which point it abruptly disappears to sea.

It can be fatal to jump to conclusions. Before we went to the Galapagos in 1964 it had seemed blindingly obvious to previous observers that a pair of frigatebirds on a nest with an egg and a free-flying juvenile persistently perching nearby, were the parents of that juvenile just beginning a new cycle. Far from it! In fact they were not the parents. The juvenile had nothing at all to do with the incubating pair. It was the still-dependent offspring of a completely different pair that had nested nearby, though the nest itself had long since disintegrated. In Abbott's case it turned out that the fork or branch to which the juvenile returned so faithfully was exactly where the nest had been. This site served for every breeding attempt, almost certainly by the same pair – vital information because the only handle on adult death-rates! If they failed to return for two or three years we could assume that they had died. And, also of vital importance, we could work out their breeding success over several years. This is crucial when you are dealing with a threatened species such as Abbott's.

One day, returning from checking the cameras, we heard wing-flaps and found a bedraggled but not seriously underweight Abbott's on the ground. Moulted wing feathers indicated an immature bird rather than a breeding adult or dependent juvenile, neither of which would have been in moult. A pinkish tinge on the bill suggested that it was an immature female; the male's bill is grey. We ringed it, avoiding red rings because that colour strongly attracts frigatebirds, perhaps because the displaying male has a scarlet throat pouch. The last thing an incoming Abbott's needs is to be attacked by one of these relentless marauders. Over the years David has picked up scores of grounded Abbott's, mostly juveniles that had made an inexpert landing in the tree. They ended up in his garden boobery on man-made, twiggy nests which he built in the bushes. Because Abbott's are so unusually and delightfully tame you can simply place them on the nest and they stay there! Most other birds would blunder off in a wild panic. So they sat there and he fed them devotedly every day. It was not easy. Where do you get the fish? Frozen fish-fingers are no good but the Chinese fishermen helped. After a while David's boobies began to fly out to sea, returning to his garden in the late afternoon just as they do in the wild. This was fascinating and surprising. It demonstrated that they had not been irrevocably imprinted on their original nest despite the fact that, in the wild, they do return to it every day with great fidelity.

*Dave Powell's 'boobery' in which he built nest platforms in his garden for juveniles that had fallen through the jungle canopy and would have died. They flew out to sea each day but were fed in the 'boobery'. Here Dave's son and 3-year old Simon (our son) feed a juvenile*

David's foster-parentage went on, in some cases, for months until, one magical day, they reached that invisible threshold and abruptly disappeared into the vastness of the Indian Ocean, exactly as they do in the wild. Who but David would have taken such trouble? Nobody except the parent boobies, and sometimes not even them.

We didn't always take the grounded juveniles home. If we knew exactly where they had come from we stood a fair chance of returning them to the nest-tree without actually climbing it with a booby tucked under an arm. A bit of applied ethology came in handy here. Experience taught us that if we simply threw the booby into the air, even from a good height, it just planed safely to earth like a paper glider. We even mounted a cross-piece on a long pole, put the booby onto it and then climbed onto the roof of the land-rover with pole-plus-booby before tossing it into the air. All to no avail. We knew that it could fly and that we had given it enough free-fall. The trouble was that it didn't want to! It was no good trying to force it. That was my clue. We had to let it reach its flight-threshold in its own good time. So we tied a crosspiece onto an extra long pole, put the booby onto it and hoisted the whole contraption upright. Then we guyed it, walked away and left the forlorn booby to get on with it. It looked pathetic, panting in the heat out there in the baking sun, but in time, just as in the wild, its flight motivation built up to that invisible threshold at which point, after the normal flight-preparation movements, it simply launched into flight. Triumph for us as well as the booby. One up for ethology!

*Returning a fallen Abbott's booby (free-flying juvenile) to its nest! By placing it high on a pole and leaving it until flight motivation has built up, it will eventually launch into flight: if thrown into the air it simply glides to earth*

Sometimes a grounded juvenile, although perfectly able to fly, sat stoically atop its pole for the best part of a day before it took off. I could easily see why. Launching into flight is one of the most dangerous things a large, web-footed seabird nesting, of all silly places, in the top of a jungle tree, can do. A slight mistake could well be fatal. If it falls beneath the canopy it is doomed because it cannot fly up from the jungle floor. One of the saddest sights imaginable is that of a perfectly healthy Abbott's booby sitting calmly on the sun-dappled jungle floor slowly starving to death. Natural selection has made sure that taking flight has become a deadly serious business. All the circumstances, very much including the booby's internal motivation have to be just right, hence the long delay before our grounded birds had reached 'take-off'.

So we took our latest casualty out to Tom's Ridge. We didn't know precisely where it had come from so the idea was to persuade it to fly out to sea so that it could then make its way back to where it should have been. In front of a sizeable crowd who had somehow gathered, perhaps to see if my theory held water, we placed it reverently on a suitable eminence. It could have been a very long wait but luckily it took off after half-an-hour.

Tom's Ridge, so lovely and remote when we first knew it, is now devastated, the jungle completely gone, but the boobies still fly over it when they come high into the island from the NW, as they have always done. So it is still the best way to get a handle on the size of the population. It isn't difficult and ought to be done regularly.

Abbott's booby is not the island's only claim to fame. We were staggered by the annual migration of the red crabs, close relatives of those that had surrounded us in the Galapagos. But Christmas crabs are land crabs and reckoned to number around 120 million though the odd twenty million shouldn't be taken too literally. There are 16 other crab species on the island but Grapsus is undoubtedly the crab. In pouring rain or high humidity they emerge from their burrows and turn the jungle floor red with their sidling forms, the air a-stir with sinister

*A 'grounded' juvenile Abbott's booby sits stoically on my shoulder before being released*

rustlings. They occupy the humble earthworm's niche, dragging fallen leaves, fruits and other debris into their burrows and, at an estimated ton of crabs to every hectare they do a powerful job. Driving along jungle tracks after rain was punctuated by a hateful crackling and squelching of shells but you simply couldn't avoid them, or their final defiant gesture – a punctured tyre. I had seven in one day.

After a while it dawns that there is not a single tiny crab to be seen. Although there, they keep well out of sight; other crabs, such as the blue crab and the yellow nipper, and even their own species, are far from averse to crab meat. Many adults meet their doom on the road as they tarry for a wee cannibalistic snack.

*Red crab*

It has long been known that the red crabs migrate from the inland plateau to breed on the seashore, but the details have only recently been filled in – by Hugh Yorkston. It is an amazing story. Mainly in November, when the rains begin, mature crabs, four or five years old, set off for the shore. They travel relentlessly but without haste. During the last tens of millions of years they have sidled unseen beneath the dense jungle canopy, as most of them still do, but nowadays there may be a railway line or a house in their path, causing a temporary pile-up but by no means stemming the flood. It is eerie to sit in a Christmas Island house after dark, listening to the persistent scratching and scrabbling of chitinous claws against walls and doors as the crabs struggle blindly toward the sea. Crabs climb on crabs. The railway line causes massive pile-ups and thousands of deaths, for they are soon cooked alive on the hot rails. Traffic on the dirt roads kills maybe a million a year; the roads turn red with squashed crabs. But compared with the natural fluctuations in the number of tiny crabs returning from the sea - some years billions, others hardly any – this death toll is insignificant.

*Robber crab*

Most of the breeding crabs migrate within a week or two, covering the distance

During their march to the sea to reproduce, the land crabs of Christmas Island may be crushed by traffic. Squashed ones are cannibalised

139

between the plateau and the shore in a few days. Males go first and spend some time on the coast, replenishing their stores of water and salt by dipping into the sea. They reach the sea-edge almost anywhere, rocky or sandy, and those unfortunates who find themselves savagely buffeted on the wickedly undercut sea-cliffs run a real risk. Unnumbered thousands die, for if a wave dislodges them they drown. Such is the price of becoming a land crab. It is safer, if slower, simply to press their back-ends into damp sand and soak up some moisture. After dipping, the males retreat to the lower terraces to dig and defend burrows in which to await the flow of females which are hard on their heels. Mating follows the merest preliminaries, hardly worth calling courtship. The male clatters his claws in disorganised fashion on the female's shell. He transfers sperm via the claws of his abdominal limbs into the female's sperm storage area to fertilise her eggs – all 100,000 of them. When the moon reaches its last quarter she lays her eggs in the sea edge at high tide. She is not alone! There may be a hundred crabs to each square metre. So massed, they produce a squealing noise which may be the result of friction between shells. Some intrepid females shed their eggs from several metres above the sea whilst clinging perilously to the vertical cliff. Since they have to rear up and shake their abdomens vigorously to detach the eggs it is something of a gravity-defying exercise which deposits many of the gymnasts into the surging sea. After depositing her eggs she may remain on the shore for a long time before the trek back to the plateau jungle.

I will never see Christmas Island again. I would find it sad and depressing now. But unless the Australian government does something monumentally stupid - by no means unknown for any government - most of the island's wonderful jungle and, above all, Abbott's booby, the very voice of the island, should be safe. I am proud to have played a major part in bringing that about. It is much better than making a million, though I wouldn't mind that as well.

*Brown boobies*

*The Treasury, Petra*

## Chapter 8

# AZRAQ, DESERT OASIS

*"To expanses of stony desert and the glories of Rum, to the Dead Sea and the Canyons of the Rift's Eastern Scarp…….. and to Petra, incomparable and fabulous"*

Guy Mountfort 1965 'Portrait of a Desert' Collins

Another bright spring dawn in the Arab village of Azraq Shishan. A cock crows, the scent of wood smoke and livestock lingers, a pleasant mix of goat's milk, chickens and a hint of donkey mingling with newly watered earth. Good, clean country smells.

We hurry down the stony slope to Ramon Said's small farm to put up our mist-nets for catching the small migrants flooding through Azraq Oasis in spring. Then we ring and release them. Ramon's work-worn wife, wide-eyed brood in tow, brings us sweet tea, the usual warm hospitality even though she has a thousand better things to do. Ramon is stocky and grizzled, in his fifties, with crow's-foot eye wrinkles and a warm smile. A few weeks later two cowards armed with cudgels and a grudge nearly killed him whilst he had his back turned. After a long spell in hospital with a fractured skull he returned gaunt and shaky but still full of spirit. I never heard what happened to his contemptible attackers.

Mist netting migrants at a staging post like Azraq is hugely exciting. You never know what may drop down to rest, drink and forage. Perhaps it's a wryneck, twisting its head round until it nearly unscrews altogether, a lovely golden oriole, a spotted cuckoo, hoopoe, handsome roller, thrush nightingale – the list goes on, a roll-call of Palearctic migrants. One day brought hundreds of yellow wagtails of several races – blue-, black- and ashy-headed together with a grey wagtail and a red-throated pipit. Oddly, although thousands of swallows and sand martins flooded through, we never saw a single house-martin. Largely on account of Azraq's marvellous bird life, we hoped it would become the jewel in the crown of the proposed Desert National Park enthusiastically supported by King Hussein. The Park was originally the brain child of Julian Huxley, 'Portrait of a Desert' Guy Mountford (1965).

Of the two separate villages at Azraq, Druze holds the darkly forbidding castle in which the Turks imprisoned Lawrence of Arabia during the 1914-18 war. Azraq Druze Arabs originally came from Syria and formed a much more tightly-knit community than the inhabitants of Azraq Shishan. In 1968 Druze school served both villages and its intense young teachers soon became our good friends. We spent fruitless hours threshing around in the morass of the Israeli/Palestinian problem. In their eyes, and absolutely rightly, Britain has a lot to answer for. The Balfour declaration, a shameless piece of cynical politics spawned by the 1914-18 war as part of the price for American aid, paved the way for Israel. War was very much in the air in the Jordan of 1968 and on March 21st Israel attacked, shattering the windows of Amman University, fifteen miles away from the guns. This particular incident soon passed but in retrospect it seems clear that the entire Azraq Desert National Park project foundered on the rock of the Israel/Palestine conflict, for that, ultimately, led to our expulsion. Azraq was merely a minor casualty but still a great pity, for it lost not only the Park but most of its superb marsh.

*Hoopoe*

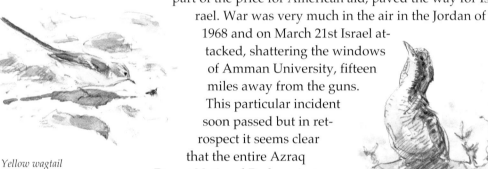

*Yellow wagtail*

*Wryneck*

At Azraq Druze we were warmly welcomed and eventually so far accepted that a young mother breast-fed her baby whilst we all sat around drinking coffee. That would not have happened in Shishan. When the locally-owned Druze bus broke down after countless punishing journeys across the boulder

*Azraq Shisham with the marsh beyond 1968. A road now runs through the village, with heavy lorries to Saudi Arabia*

*Black-winged stilt at the edge of the marsh*

and basalt-strewn desert, the village men set to work and stripped and rebuilt it. That, too, would not have happened in the much less-cohesive Shishan. Just outside the village a market garden grew barley, beans, alfalfa, almonds, tomatoes and other vegetables in amazing contrast to the arid surroundings. The only difference; the vegetable garden got some water. Again, a communal effort.

Azraq's water and pervasive greenery was a magnet for migrants. It also attracted village boys with catapults and malign intent, though to them just fun. Red-throated pipits, yellow wagtails, wood-sandpipers, and dozens of other species fell victim. We rescued a lesser bittern whose wing their 'fun' had shattered, and took it home. It caught small fish in our shower-well with amazing dexterity before retiring to roost, neck and bill held vertically, alongside a broom handle – the nearest thing to a reed-stem! Sometimes it roosted next to the lavatory pot, an uncomfortable companion with its dagger bill and unblinking eye fixed on your backside. Hunting is a mania in Arab culture. Cruelty and the beauty of wildlife seem to mean nothing. The local customs officer was all for testing his marksmanship on a magnificent black-eared wheatear in full breeding plumage. To him it meant no more than a fly- which, come to think of it, may be deeply philosophical, though I doubt that he thought of it in that way. In downtown Amman a sweet old lady smiled approvingly as a small boy shot sparrows with his catapult. I saw hunters take pot-shots at hoopoes, just for fun. It is simply a different attitude than ours and it runs deep. Our sentimentality seems crazy to them. But then it is not long since we staged cock-fights and baited chained bears with dogs. And thousands still hunt for fun rather than necessity. We are a mixed-up lot in more ways than one.

I lost count of the number of times we drove across the desert from Azraq to Zerka. It was always an adventure for at that time there was no road and no set route. Where you went depended on the vehicle's clearance, its decrepitude (usually extreme), the weather and the driver's fatalism. It was always a choice between mud and water or boulders and dust. Arab drivers, attempting the impossible and failing, could always blame poor old Allah. It is the will of Allah, not my bloody stupidity. The route crossed a 10km stretch of a sandy, hard-packed, pancake-flat Qua or Playa. Years later it grabbed world headlines when hi-jackers forced a Pan-Am passenger plane to land there. It was a darned sight better than many a runway! In dry weather with a strong wind blowing, vehicles crossing the Qua moved inside an impenetrable cloud of dust. It was a game of Russian roulette. You could hear the roar of an approaching Goliath but what with your own dust and the cloud thrown up by the approaching colossus you couldn't see anything. Luck and the wide expanse of level sand which allowed frantic last-second evasion was all that averted head-on collision. It was petrifying. After heavy rain the Qua flooded to a depth of around eighteen inches, which was ideal for black-winged stilts but not so good for cars. The trick was to forge steadily across this vast shallow lake without stopping, which would have sucked water back into the exhaust. After torrential rain vast areas of the desert became flooded, including patches which nomadic Bedouin had roughly ridged and planted for a 'catch-crop'. Negotiating one of these 'gardens' was sheer luck and the wit to keep going. One proud day we were the only ones to make it from Zerka to Azraq and the watermark reached half-way up the land rover doors. All that is now a thing of the past, thanks or otherwise to a new tarmac road.

After a time our growing menagerie further complicated our trips to Amman. The old land rover had to accommodate us, perhaps a passenger or two, three pi-dogs, two jackal cubs, a wolf cub and several jerboas. It was not a happy mix on a long, hot, jolting and dusty journey.

*Our bedouin dogs*

*Rescued wolf cub (top left) forepaw badly injured by a trap, with a rescued jackal cub*

*Our wolf and jackal (top right) finally settled in Dudley zoo*

*Jackal cub indicating submission (ears back) to bedouin bitch*

How on earth did we accumulate this menagerie? Well, the abandoned pi dogs would have been killed had we not rescued them. The wolf cub had been snared in a wire noose which cut deeply into its 'ankle' and by the time we found it a suppurating wound had developed above a grotesquely swollen foot. The two jackal cubs, dug out of their den by locals, had been tethered on the salt pans. The salt had eaten deeply into their pads and the poor wee beasts could only hobble painfully, albeit hilariously as they slid and slithered when the whole menagerie chased each other across the polished floor in our Amman flat. Before leaving Jordan, we shipped the wolf and jackals to Dudley zoo as they would surely have died had we released them in Jordan.

How did a seabird man ever become involved with the desert? How indeed! When the mixed flocks of boobies and frigatebirds escorted us away from Christmas Island, glad to see the back of us, we hadn't the foggiest notion of working in the Jordanian or any other desert. Like much else at that time it was Harold Wilson's fault. In March 1968 the Labour government, once again short of cash, froze all new appointments to universities. My intended lectureship at Aberdeen was one of the minor casualties.

Luckily for me, a recent expedition to Jordan by Guy Mountfort, Julian Huxley and Max Nicholson had put the spotlight on deserts. Mountfort's 'Portrait of a Desert' had shown Azraq to be a major crossroads for migrants between Africa and Eurasia so the idea of creating a desert national park centred there took root, supported by King Hussein and by the International Biological Programme of that time. I have no idea why Max invited me to initiate the programme. Maybe, knowing something about our shoestring Galapagos and Christmas Island expeditions, and that we were free and with no children to complicate matters, he thought we might make a silk purse out of a camel's ear. Whatever the reason, it was a godsend for us, so off we went.

A few weeks after we arrived in 1968 Max, and Morton Boyd of the Scottish Nature Conservancy, flew out to help us put together a proper Constitution for the Station, embodying its role within Jordan and defining its funding. There was also the little matter of Trustees and a 'boss' for me. Everybody has to have a boss! In fact, this totem figure never materialised though in theory a Council of seven members, including representatives from Amman University, The Royal Society for the Conservation of Nature (Jordan) and the Ministries of Agriculture and Tourism should have overseen my work. Like a lot of these things, it looked good on paper but it didn't work because there was no single person to chair or drive it forward and I'm not an administrator. These details were meat and drink to Max who had, after all, been the architect of our own highly successful Nature Conservancy in the 1950's. He was a familiar and somewhat dreaded figure in Whitehall's corridors of power. Long afterwards it occurred to me that it may well have been because he was such a gadfly that he never received the honours he richly deserved. He should certainly have been knighted. Fred Holliday, who later, as he would fully agree, played a much less pivotal role in the Conservancy, received that recognition. The day after Max and Morton arrived at Azraq, he hammered out a constitution for the station and June gallantly typed it on our battered old portable. There were no lap-tops or other sophisticated gadgets in those blessedly simple days.

Before Max departed – I nearly wrote 'fled'- Anis Mouasher, a Jordanian businessman and a major patron of the Royal Society for the Conservation of Nature (Jordan), threw a party for us and some Jordanian notables. Max revelled in such occasions and was on top form, expatiating loudly in his unmistakable, spittle-laden tones, on what he called 'the three harmonies', between people and people, people and environment, and people and government. Is there anything there for our current lot?

We first flew into Amman just before a wonderful desert sunset, and were instantly engulfed in the usual noisy, gesticulating airport scrum. Our University contact failed to meet us, having read '19.00 hrs.' to mean 9.0 pm. But a kindly Jordanian taxi driver, surely the most trustworthy representative in the world of his cut-throat trade, took us under his battered Mercedes wing and delivered us to the 'Hotel Select' on Jebel Lwebdi, one of Amman's famous seven Jebels or hills which give the city its unique character. Soon afterwards our abashed Jordanian friend hastened up and, Arab hospitality at stake, insisted on taking us out for a meal, which was slightly spoiled by armed Fatah guerillas lounging around and collecting 'donations' from the diners. Later, King Hussein threw out all such insurgents but in early 1968 they were a menace in Amman.

Next day we called at the British Embassy which liked to know where people were and what they were up to. And we paid an unplanned visit to a Jordanian police cell. It all happened in a flash. A couple of Iraqi soldiers spotted us wandering around in the centre of Amman, trying to memorise landmarks as one does in a strange city of which one expects to see a lot in the immediate future. Understandably, they became suspicious. What were we up to, stopping every few minutes to gaze around and point. They accosted us in Arabic and of course we couldn't explain ourselves. I doubt if it would have made much sense to them even if we had been able to speak Arabic. Setting up a centre to study desert wildlife? A likely tale. A noisy and excited crowd soon gathered, all shouting at once, and we were arrested and hauled off to a grossly overcrowded police cell. Our initial alarm was quickly dispelled by a charming, elderly Jordanian – I have no idea what heinous crime he had committed – who literally wept with shame at this violation of Arab hospitality. "I'm sorry, I'm sorry" he repeated over and over again. Somehow I can't quite imagine one of us reacting like that to an incarcerated Arab. The police demanded our passports which we had left in the hotel where, by sheer chance, Subhi Qasem, our contact from Amman University, had gone to see us, only to find a shouting match between police demanding to search our room and hotel staff stoutly resisting. He quickly sorted it out but it could easily have turned nasty. Some time later a British tourist was shot and very nearly killed by an over-zealous Iraqi soldier in the centre of Amman and eventually the Iraqi military threw us out of Azraq. Poor old Iraq.

Amman University provided one of our two most important links in Jordan. With its support we had great hopes of initiating a wide-ranging research programme which would attract American and European post-graduate students with their all-important grants. Azraq had huge potential and we drew up a long list of possible subjects for research with the aim of circulating it to hundreds of University departments. Dreams, mere dreams. We needed Jordanian or other Arab academics to agree to supervise such students and it was Amman University, not me, who had to provide the names of people able and willing to do this. Alas, it never happened. However explicitly I made this clear, Amman University still seemed to think I should come up with the list of names!

A huge fenced area covered thousands of acres at a place called Shaumari, a few kilometres south of Azraq. It had great potential for research and for breeding native oryx, Arabian gazelles, houbara bustards and ostriches, shielded from persecution and from overgrazing by goats, sheep and camels which had utterly degraded the greater Azraq area. Amazingly, there was even a large, complex laboratory belonging to the Ministry of Agriculture and built as part of an American-funded project to investigate the position and extent of the underground aquifers or water sources of the area. I shudder to think what that had cost. On paper it all seemed wildly exciting, with great potential but it came to nothing. Many years later, long after the Azraq project had foundered, the World Wildlife Fund and the International Union for the Conservation of Nature actually did support an oryx breeding programme at Shaumari. Four male Arabian oryx were flown from San Diego zoo, followed by four females, another male and two females from Qatar. The first calf was born in 1979 and eventually 31 oryx were released into the wider Shaumari area. Half a dozen ostriches

*Oryx were bred in the Shaumari Reserve near Azraq for eventual release into the desert (they used to be abundant)*

which used to be common here but had long since been wiped out, were introduced in 1985 but by then Azraq itself had been sadly spoilt.

Our high hopes – in fact our firm expectation – that the biologists at Amman University would enthusiastically embrace the various research opportunities at Azraq faded like dawn mist. The academic botanists and zoologists showed scant interest, which I simply couldn't understand until it dawned on me that Arab university tradition had always been very different from that of the Western world. Fieldwork and Natural History, of low esteem and rather below the dignity of an Academic when compared with laboratory or literary studies, belonged to a different value system. So we just got on with our own observations: a great treat. We loved Azraq, but it was a missed opportunity for Jordanian academics.

From March to July the tide of migrants and the birds of the Azraq marshes and surrounding desert kept us happy, but we had hoped to do much more, both at Azraq and at Amman University. In truth we had to spend far too much time on practicalities. The promised transport turned out to mean merely that we could hire a land rover and driver from the Ministry of Agriculture, which would rapidly have swallowed up our IBP grant. And problems loomed on the accommodation front. Max Nicholson's simple advice, derived from his personal experience of Whitehall (and perhaps this explains why he never got that knighthood): "go and camp on the University's doorstep until, to get rid of you, they provide a land rover". A fat lot of good that would have done. The university didn't even own a suitable vehicle nor the cash to buy one, and most assuredly not for me. My much more practical solution, born of my 'do it yourself' philosophy, involved flying home and driving our own land rover back out. We had to find somebody who would stand surety for a sum equal to the value of our vehicle, payable, if need be, to the Jordanian government. Aberdeen University, on the word of Prof. Wynne-Edwards, kindly shouldered that potential burden. So for us back to Aberdeenshire to collect out three-year-old land rover from its rusting place and then tackle the long haul across Europe to Jordan.

We sailed from Marseilles to Beirut calling in at Naples where it rained in buckets making the impossibly steep stone-paved alleyways lethally slippery. Ancient trades such as leatherwork, sewing, embroidery and shoe-repairs still flourished in door-ways and dim recesses. Household slops cascaded from high, shuttered windows in the narrow canyons. The violent downpour soaked us and shrouded the old harbour far below, guarded by its monstrous Bastille. Seedy Alexandria failed to entice us ashore and marked the point at which, ominously, the ship's carpets were lifted, leaving the bare boards before the influx of Arab passengers. Then it was Beirut and the usual and dreaded stress of Customs control, heat, noise, agitated travellers and the embarrasing language barrier which always made us feel so useless. Oh to be a lin-guist, however bad.

Faithful Abdullah from the RSCN(J) met us and drove our land rover to Jordan through Syria, at that time barred to the British. We flew to Amman. There is some-thing menacingly obtuse about officialdom, a sort of collective stupidity like a spider's web, which nobody can break free of. We went through the hands of no fewer than twelve minor officials, taking two solid hours, just to obtain a form entitling me to col-lect my own vehicle for which my International Carnet should in any case have been sufficient since it guaranteed Jordan against loss. What had been the point of all the hassle involved in getting it? But we triumphed. We were the only ones to succeed in freeing a vehicle that day from the compound, overflowing with a depressing collec-tion of cars, thick with dust. The guardian of that unhappy lot was visibly surprised when we presented the magic form. From Aberdeen to Azraq in one far-from easy move, and back to the birds.

Was Jordan delighted that this idiotic foreigner was prepared to thrash his own land rover for 15 months in the desert on their behalf? It certainly didn't feel like it. Every three months I had to go cap-in-hand to some Ministry or other to renew the permit which graciously allowed me to punish my own vehicle in their service and with nothing paid for wear and tear or fuel. Hordes of shouting, gesticulating Jordani-ans besieged the imperturbable official sitting at his paper-strewn desk. At intervals a curt command sent a waiting minion scurrying off clutching a sheaf of forms, to ob-tain the signature of an even mightier deity behind a closed door. Occasionally a def-erential cup of black coffee appeared on his altar. Meanwhile I stood there sticking out like a blond 6' 2" sore thumb, waiting for eye-contact so that I could present my peti-tion. Not a word did I utter until spoken to. I knew my place and in Allah's good time it worked.

At first we lived in the old shooting lodge, an unpretentious but solid structure built by the British as a field hospital in World War 1. Electric power came from an old Lister generator built in Birmingham around 1900. This venerable machine thumped away, lovingly tended by an old Russian with a deeply seamed face, a well-worn khaki shirt and a black beret which never left his head. Punctually at six o'clock every morning he brought tea, bread and jam to our spartan room. The lodge belonged to the RSCN(J), a new body which was in fact the old Jordanian Hunting Club!. Every weekend parties of jovial Bacchanalian duck-shooters assembled here, their car-boots stuffed with good cheer nominally denied to Muslims. The Society 'administered'

Azraq and it was through their generosity that, albeit temporarily, we were allowed to live in the lodge. That, alas, was soon to end.

By August we had moved into a newly-built but unfurnished aluminium hut. I'm not one to worry about status but I did begin to wonder exactly why people from UNESCO and US Aid had it so cushy. It is not in the least surprising that these organisations spend millions when they provide their personnel with a house, a four-track vehicle and a servant. A servant! We came out with nothing, provided our own transport and had to find, bargain-for and carry everything we needed across 60 kilometres of trackless desert, all extra stress getting in the way of our work. Stores had to be listed, bought in Amman and, if fresh, kept cool until we could take them out to Azraq. Every possible snag had to be thought out beforehand for nobody told us anything. When we did get back to Azraq, hot, dusty and tired only to find the Lodge deserted and locked, with all our personal stuff inside and no note about the key or when anybody would come and open up and no way of finding out, it was way past time for a primal scream. Everything seemed to happen ad hoc, on-the-hoof. It may or may not have been lovable but it was not efficient.

*The aluminium chalets which the Royal Society for Conservation of Nature (Jordan) built at Azraq Shislan in 1969*

Azraq life proved always exciting, quite apart from the wonderful bird-life. One day, great activity up at the Bedouin-manned Desert Police Post with its racing camels tethered outside, heralded the arrival of Wasfi-el-Tal, a former Prime Minister of Jordan and a most delightful man. He was assassinated in 1971 in Cairo by four of the Black September members of the PLO. His assassins left prison a mere two years later. A feast, or mansef, was prepared in the great ceremonial tent strewn with Bedouin rugs and cushions, all familiar enough to Wasfi whose own grandfather still lived in a tent. King Hussein's grandfather famously preferred his tent in the Palace grounds to the Palace itself. Jordanians are much closer to their nomadic desert roots than the urbane Syrians, Egyptians or Palestinians and, perhaps with good reason, consider themselves altogether hardier than these. As a boy Wasfi was expected to be self-reliant in a way that would horrify today's parents and drive social workers into outraged orbit. Food at home was simple and occasionally he was sent out into the desert to fend for himself, living rough and shooting and cooking his own food.

Wasfi-el Tal's immaculate white house, approached along a tree-lined drive and through graceful arches, sat on the shoulder of one of several barren hills, its terraced garden, lovingly created over many years, full of small oaks, pines and flowers. From his political library I borrowed King Hussein's 'Uneasy lies the Head', a tale of immense courage and dignity in the face of politically-motivated lies and hate-propaganda. Of Wasfi's own tales the only one I can recall involved a UN meeting at which a minor African dignitary passionately denounced his chief for eating one of his wives. Our own Azraq home, an unlined aluminium shell in a row of others, perched on the slope just below the Lodge and overlooking Azraq Shishan. They were built by the RSCN(J) though don't ask me why they chose aluminium, baking hot in summer and freezing in winter, the exact opposite of the thick-walled old Lodge. Our shell may have been thermally challenged but its elevated position, gazing over the village and the green sweep of the marsh beyond more than compensated. What did it matter that we sat on upturned petrol cans and ate off concrete blocks?

Trains of heavily-laden camels periodically crossed the heat-hazed desert with its vivid mirages of water and cliffs. These summer visitors arrived when the fringes of the Qua had dried out. Their encampments mushroomed where, a few weeks earlier, garganey had dabbled and black-winged stilts waded on match-stick red legs. When the last of the water finally evaporated it left untold millions of small fish (Tilapia) stranded. Some of the visiting Bedou came from Yemen, others from Saudi Arabia. For two or three pleasant months they grazed their animals on the feathery tamarisk and spiny shrubs; camels would happily browse on barbed wire for all the difference the lethal spines seemed to make. All these hundreds of animals greatly exacerbated the already severe grazing pressure around Azraq. Then, when the sandy Qua began to darken and soften from below, hinting that it would soon flood again, the nomads packed up and the camel trains moved off once more. Each evening, scores of camels, urged on by the drover's ringing cries, came to drink at the Azraq pools. They splayed their awkwardly long legs and sucked up gallons of water at one go. As they lifted their heads and the water began its long trek down their throats, pendulous upper lips wobbled and they urinated copiously. Water in, urine out.

*Bedouin camels drinking at Azraq Shishan pools. Pure water flows under the basalt from Syria and emerges at Azraq*

*Fishing in the Azraq pools using a hand-thrown net or shebak*

Your true Bedouin is half man, half camel. Can any man be more firmly wedded to an animal than the Bedou to his camel? Not the Argentinian gaucho to his horse nor the Tibetan to his yak. It is the Bedou's transport, food, shelter, warmth and, often, his very life-line. Camels are not just camels; that is an ignorant outsider's word, my word and, I daresay, your word. A camel can be any one of 600 animals depending on its age and condition. Six hundred words just for a camel! His skills with the animal define a Bedouin male. Goats, sheep and donkeys are for women and children. Besides its economic value there are the close bonds forged by hazardous desert marathons when camels alone save the Bedouin from a lonely death in the burning wastes; a riding camel can cover more than 100km. in a single day. We met a young Bedouin who, alone with his camel, had journeyed for 28 days across arid wastes. Even Wilfred Thesiger's dauntless young Bedouin friends would be hard pressed to beat that. Think of the water, kindling and food he must have carried though there would not be much of the latter – maybe a bag of oats, some dates, sugar and tea and perhaps the odd hare if his ancient rifle managed to bring one down. "To the pragmatic western mind nothing is more confusing than the complex character of the Bedouin, with his innate courtesy and thoughtless cruelty, his passionate belief in personal liberty and subborn resistance to change, his scorn for material possessions and contempt for life" (Guy Mountfort, 1965)

A prosperous caravan on the move is quite a sight. One crossed Azraq Qua in the golden glow of sunset, heading south, maybe to Saudi. About 70 laden camels, some with supple black waterskins, fleeces, fat bundles of faggots, cooking pots and all the other accoutrements of a self-sufficient clan on the move, and not a single wheel to be seen. A string of unladen animals and scores of sheep, goats and their inevitable canine outriders trotted behind. The richly caparisoned lead animal carried a howdah and two ornate rifles, resplendent with mother-of-pearl butts, slung at its side. Apart from the rifles nothing had changed since pre-biblical times. I find it hard to believe, given all that has happened since, but in that year (1968) there still existed desert Bedouin who had never, in their whole lives, seen a newspaper, heard a radio or knew what a telephone was. Out of 26 Bedouin interviewed for a Jordanian magazine, nine had never heard of a cinema and as for the radio —"Radio? Walahee (by Allah), I am listening to riddles". By now, doubtless, they all have lap-tops, mobile phones and I-pods - more than I have. The Arab Bedouin's way of life was entirely different from and far superior to that of other camel-using nomads, such as the Ghiljai of Pakistan who migrate every year between the Indus and Afghanistan to trade textiles, clothes, shoes etc. which they buy during long sojourns in cities. That was not the Jordanian Arab Bedouin's way. The desert was his home and the only one he wanted.

Real Bedou are proud people, valuing oral tradition, desert lore and, above all else, their concept of personal honour. In the absence of written records, which no nomadic society could possibly carry around, they have developed phenomenal memories. National geographic boundaries and interests have never meant anything to them which has caused endless trouble with today's fiercely political states. Bedouin did not belong to a country such as Syria, Jordan, Egypt or Iraq, with all their messy politics and shifting allegiances. Their tribe alone claimed their absolute loyalty. Exclusion from it, the ultimate penalty for a serious crime, was tantamount to a death sentence for you cannot long survive alone in the desert. To us modern Europeans, and even more so to Americans, pampered, flabby and avid for all the pleasures we can wring from life, the Bedouin's lot seems incredibly harsh, one long punishment. At its end, usually at a pretty early age, he is laid to rest in a simple desert grave marked by a stone or two. But those who have spent even a little time in his haunts can gain some slight inkling of how he comes to love the desert and his freedom. In the cool of the evening he returns to the great striped tents. The camels are stockaded, tethered or left to wander. Then, usually around sunset, whilst the women busy themselves behind the tent, making the evening meal, one of the men begins pounding the roasted coffee beans in a wooden mortar — the drumbeat can be heard a kilometre from the camp. Whilst the water comes to a simmer the invitation goes out to all who

can hear it to come, join us. This is the epitome of desert hospitality, simple and strong. The coffee, a thin, almost greenish liquid, a clear coffee cardamom distillate, takes a long time at a slow simmer. Bedouin coffee is sharp and uncloying, stimulating to take in tiny sips. Beans are roasted in a long handled metal spoon held in the ashes of the fire which always smoulders in the tent.

The water which filled Azraq Shishan's crystal pools and sustained the bird-filled marsh beyond, began its journey in the Jebel Druze of Syria. It flowed unseen beneath the blistering black basalt until that savage rock petered out at Azraq. These pools, the heart of the Desert National Park, have

*Bedouin woman roasting coffee beans*

been oases of rest and healing for generations of nomadic Bedouin. Thousands of years before them, at the dawn of human history, stone-age man roamed this rich and beautiful area. These early people left thousands of flint tools and chippings near an attractive freshwater spring –Lion's Spring- where I picked up a fine hand-axe lying on the ground as though carelessly left there the day before yesterday. In those far-off days gazelle, oryx, wild asses, wolves, jackals, hyaenas, lions, leopards, cheetahs, ostriches, bustards and who knows what else roamed the desert. But by the late 1960s millions of gallons of water – far and away too much – were being pumped from

*Bedouin offering coffee to June*

Azraq and Irbid to Amman. The pools and marshes, still pristine in 1968-9 have since begun to dry up and with their demise would go the habitat for the thousands of wildfowl and waders that depend on the marsh and the flooded Qua. But in 1968 that unhappy picture was still to unfold and for us, the emerald green marsh dotted with islands still swept away into the distance until it met the golden sand and the mauve and black hills beyond. In the pools naked brown Bedouin boys splashed happily whilst, completely separated, an occasional girl, garbed from head to foot, tried to soak herself, a miserably unfair division of glee enforced, like so much unfairness, by religion. Islam, perhaps even more than Judaism and Christianity, is grossly male-centred.

It was the unexpected that made Azraq so exciting. Lying at the migratory crossroads between Africa and Eurasia the wide variety of its birdlife constantly surprised us and we often found it hard to believe that we were in a desert. One day hundreds

*Reeds harvested from the marsh*

of coot bobbed around in a flooded area where yesterday there had been none, as though they had merely toddled from one park-lake to another whereas in reality they had crossed many miles of hostile desert. Where on earth had they come from? Had they flown in as a flock? Or the surprise might be a huge Caspian tern, some black terns or once, amazingly, two Arctic terns, the ultimate ocean wanderer which migrates yearly from the Antarctic to breed in the Arctic, one of the longest annual migrations of any bird. I don't blame anybody for thinking they must have been common terns. Even that would be amazing. In early April in an area of sparse Phragmites little egrets fed alongside black-winged stilts when a small flock of gull-billed terns flew in. The air stirred with teal, shoveler and garganey, marsh frogs swarmed astronomically and the water seethed with Tilapia which in turn attracted harmless black water snakes. Water rails squealed but, as ever, were nowhere to be seen. The reeds concealed nests of lesser bitterns and purple herons. All this in the middle of the desert. To cross the marsh took two hours, wading through waist-high water, squelching through sulphurous ooze, squirting catfish underfoot. Given the uncertain temper of water buffaloes it was scary to emerge onto one of the many small islands and find one staring at you. It seems they were sold 'on the hoof', to be shot, but I never discovered how the carcasses were dragged out of the marsh – no easy task. They may have been butchered in situ, which would make more sense.

*Bonellis eagle*

*Rough-legged buzzard*

*Sakr falcon chasing a sandgrouse*

Azraq hosted spectacular migrants, especially the raptors. Eagles, buzzards, harriers, vultures, hawks, falcons and ospreys kept one constantly amazed. A peregrine in hot pursuit of a towering redshank, desperate to keep above that deadly stoop; a merlin skilfully outflying a sand martin; the dusky beauty of a male red-footed falcon; a ginger-trousered hobby or the winnowing flight of lesser kestrels as they poured across the desert. In many countries such a feast would attract hordes of eager bird-watchers; at Azraq, blessedly from our selfish viewpoint, not a soul.

Azraq's larks, too, came in astonishing variety. My favourites were the leggy hoopoe larks with long, down-curved bills and white wing-flashes, and Temminck's horned larks with startling faces and upturned feather tufts or horns. But for sheer hardiness pride of place had to go to the drab desert lark which somehow managed to survive the brutal heat and searing aridity of the black basalt boulder plains – hell on earth. What did they find to eat?

Birds were not the only winged migrants. Tens of millions of butterflies, especially painted ladies and red admirals fluttered over the springtime desert, feeding on the nectar from the carpet of flowers "when the desert blossoms as the rose". It didn't last long.

*Hoopoe lark display flight*

*Desert horned lark*

The gentle little jerboa, supreme among desert-adapted mammals and never to be confused with the vulgar gerbil, must be one of the most captivating mammals on earth. It resembles a miniature kangaroo but is much better-looking, with a blunt nose which it uses for tamping earth plugs into the mouth of its burrow, to seal in the moisture, huge eyes for nocturnal foraging and enormous sensitive ears which help it to evade desert foxes, jackals and owls. Its long, powerful hind legs propel it in gigantic leaps with erratic changes of direction to thwart a pursuer. The long tufted tail provides a sensitive balancing pole and rudder. Jerboas emerge at night, when it is cooler, and they escape the lethal heat of the desert day within their burrows. They seldom drink and can live indefinitely on dry seeds. Oddly enough they had never been adequately studied and I planned to take a few back to Aberdeen, breed them in a huge enclosure and study their social behaviour in detail. But it isn't easy to catch jerboas, as any fox will tell you, for their whole life is one long escape from predators. They do not jump along in a straight line but rocket off at right angles just when you are about to pounce. We found that the best way to catch them was by night-time chase across the desert in a land rover. June sat on the bonnet with a butterfly net and at just the right instant dived off and with a deft scoop ensnared the poor little jerboa – or more likely, missed by a mile. Amazingly this injured neither woman nor beast and we actually caught some.

Like many animals with meagre food, jerboas have evolved several life-saving strategies. Take breeding, for example. Unusually for a small rodent they take several years to become sexually mature. Even then they breed sparingly, producing very few young and only at long intervals. All this is the exact opposite of the system in rats,

*Powerful elongated hind legs of jerboa enable it to leap erratically to elude a pursuer. Hair–fringed hind feet provide purchase on loose sand*

rabbits, mice and voles. A brown rat can breed when little more than a month old and during its lifetime it out-produces a jerboa by hundreds to one. Before we left Jordan I sent a few jerboas for our friend John Busby to look after until our return and they actually bred – probably the first time this has happened in captivity. I planned to build an enclosure for them but alas I merely proved the aptness of Robbie Burns' line and "gang agley" my best laid plans certainly did. Long before I cut the first sod for my jerbarium a rat got into their temporary enclosure and ripped their throats out. It was sheer class hatred, the yob and the gentleman.

Despite our meagre achievements in Jordan we loved our time there, in no small measure due to Abu Talal, the Director of the RSCN(J). He was the very essence of Arab hospitality and took it upon himself to show us the real Jordan. So, one grey dawn in May 1969 two land rovers set out from Amman laden with food and

*Jerboa*

camping gear. One was ours and the other carried Abu Talal and two miserable drivers who knew only too well what lay ahead. At our first pit stop, a newly-drilled borehole in the desert, who should we find shaving in a modest little hut littered with battered old beds and a few sticks of undistinguished furniture but Sheik Faisal, MP. Casual he may have seemed but don't be misled. Tribal fealty is so strong in Bedouin culture that there were many who would not have hesitated to kill at a word from him. That, at any rate, was what Abu Talal told us. After a few pleasantries in Arabic, which of course passed unimpeded between our uneducated ears, and the inevitable coffee, it was back to the King's Highway and then south to the amazing Wadi Rum, at that time innocent of any road, just as in the days of Lawrence of Arabia. We camped near the foot of mighty Jebel Rum against which even the lammergeyer's vast wingspan shrinks to that of a pigeon. The floor of this immense wadi is mainly flat but broken by jagged peaks. Here squats the famous Desert Police Post and a scattering of Bedouin tents. The men, in flowing khaki dress and red-checked kafirs (headdresses), the black rope agal set precisely at that jaunty angle which no foreigner can quite copy, sat around drinking sweet tea. At sunset choughs flew and called high against the rosy-fissured cliffs. After nightfall the silence was carved out of black velvet. The clear starlight shrank the width of the wadi, setting the far cliffs a mere few metres away though in truth they were over a kilometre distant. A nearby mole-hill turned into a distant Bedouin tent. During an utterly tranquil night in the open I awoke to see, too perfect to be real, a fragile, silver-sickle moon poised on the very peak of a great black crag. Morning revealed the tracks of striped hyaenas around our camp beds.

In those days Aquaba was a dusty Red Sea fishing village with a phosphate factory. Abu Talal, may his tribe endure, knew of a good fish-restaurant, a modest green wooden shack, which, after an interval long enough for an Arab dhow to have gone out and caught the fish, produced a superb meal. This was entirely Abu Talal's magic;

**161**

*Tents of the Beni-Sakhir bedouins and the polide post beneath the 1000' Jebel Rhum 1969. At that time there was no road through Wadi Rhum*

they would never have even entertained the idea for us. The place wasn't even officially open. Rank counts every time with Arabs and Abu Talal was a natural autocrat with a superb air of disdain when he wanted. When I next saw Aquaba more than twenty years later the Iran/Iraq war which sucked in Jordan on Iraq's side had transformed it into a tarmac desert, parking-lots for tanks and army vehicles. It had an airport, a sea-port, new hotels and business-cum-shopping areas, a peaceful Arab backwater dragged into 20th century mayhem. But my abiding and most painful memory is of a forlorn sheep, head hanging low, tethered on the dusty pavement in the brutal sun, waiting for the butcher's knife. Nobody gave it a second glance; it might as well have been scrap-iron.

After Aquaba Abu Talal's odyssey took in Wadi Musa's Rest House and a tranquil evening sitting under the branches of a luxuriant fig tree. The poor old drivers, their early forebodings fully realised, spent a cold and comfortless night in the laden land rover and awoke next morning in a foul and mutinous mood, but, though sullenly, they drove us to Petra. With Patrician disregard, Abu Talal serenely ignored their temper. At that time Petra boasted only a single Rest House which blended discreetly into the background. A giant cave with rectangular recesses carved out of the rock and pleasantly lit by traditional Arab brass lamps with Hebron glass bead shades, served as the dining room.

Petra! Ah, Petra, now the darling of the Jordanian tourist trade. Remote, beautiful, mysterious Petra, created in biblical Edom now called Jordan. In the 7th century BC a semitic tribe from southern Arabia, the Nabateans, regularly raided the great caravans which moved between Arabia and the Medditeranean. Gradually this troublesome

162

tribe became wealthy, put down roots and supplanted the Edomites in southern Jordan. By now well-established in Northern Arabia too, they turned from poaching to gamekeeping and, for a consideration, began to protect caravans instead of raiding them. At Petra they built temples, cut tombs out of rock-faces and built aquaducts to carry water from the surrounding hills into the city. So the old Edomite capital originally called "Sela" became "Petra". At one time the Nabateans controlled land from the Red Sea to Damascus and occupied all of Sinai but at the height of their power they had the bad luck to run up against the Romans and in 100 AD Nabatea was absorbed into the Roman Empire. Many of Petra's present-day remains are Roman. A few centuries afterwards Petra seems, somewhat mysteriously, to have declined and was utterly lost to the Western world -gone without trace- between the 12th century and its accidental and most remarkable re-discovery in 1812.

The brochures, for once accurately, describe Petra as one of the world's wonders. It is approached through a narrow canyon that jealously guards its secret until the final bend. Then, glowing pink and red in the early morning sun and throwing brilliantly black shadows, the Treasury magically appears like a cosmic conjuror's trick. It is massive, intricate and, unbelievably given its gigantic proportions, carved out of solid rock. Carved! How on earth did they do it with primitive tools and simple muscle-power. On our first visit Petra was a-glow with purple oleander. Sheep grazed here and there whilst their shepherds perched on the surrounding crags. We climbed high on steep, worn steps to El Deir, the monastery, whose great dome gazes clear across majestic Wadi Araba to Israel. At the time of our first visit a simple catch-crop was being harvested in the old way, by hand. And not another visitor in the whole place. Perfect. Groups, shepherded by vociferous guides in full multi-lingual flow, do rather spoil things.

Abu Talal still hadn't quite emptied his magic sack, so we returned to Wadi Musa and then, laden with fresh mutton and other provender, carried on to Shobek, stopping off at a roadside pistachio tree, obviously a well-kent landmark for in its shade rested Anis Mouasher and party. Shish kebabs and arak went down well before

*The worn approach to El Deir, the monastery at Petra*   *The monastery; behind lies Wadi Araba and Israel*

carrying on to Wadi Dhana in mountainous southern Jordan, the last refuge of some larger predators, perhaps panther and certainly wildcat, jackal and hyaena. Then to Karac's crusader castle and Wadi Mujib's hair-pin bends snaking down one side and up the other of this terrific chasm. At Mount Nebo, a fine mosaic floor depicts birds and mammals now, alas, extinct. From here Moses gazed into the promised land which he never entered but which his descendants have done with a vengeance. Religion again.

It would take several books to do justice to Jordan's archaeological treasures. One of its finest, the Roman city of Jerash, has a marvellous street of columns still marked by the ruts of charriot wheels, and a classical, elliptical forum. In 1968 it was still largely unexcavated, with artefacts lying everywhere. Whilst this undoubtedly lent an air of authentic antiquity and neglect, it seemed perilously trusting.

*The ruined Roman city of Jerash*

Back at Azraq, sadly and very prematurely, everything shuddered to a jarring halt. We had several modest projects under way but it was not to be. A polite knock at the door of our hut presaged a sombre and slightly embarassed Fayk Wazani, our friend from the village, accompanied by a polite but immovable Iraqi army officer. Why he needed Fayk I don't know, for in faultless English he gave us 24 hours to pack-up and get out. In vain our protests, useless our official permits which the Jordanian army had always honoured on the occasions when our paths had crossed in the desert. He merely said, though with the hint of a smile, "I don't care if King Hussein himself gave you permission, for your own safety I am telling you to leave. This is now a military zone". We knew a little about the Iraqi army and so we went.

Even the journey home was not simple. As 'persona non grata' in Syria we could not drive through to Beirut so, as on our outward journey, the invaluable Abdullah of the RSCN(J) drove our laden land rover from Amman whilst we flew out to meet him.

We had made an enormous wooden box to fit on the roof-rack and filled this and the back of the land rover with the odds and ends collected during our time in Jordan. Most of it was quite useless, none more so than the huge jars of olives which soon turned rancid, doubtless resentful of the heat and jolting, but 40 years on we still use the striped Bedouin rugs which Abu Talal negotiated for us . And we treasure the huge engraved copper plate, a metre across, once used for ceremonial mansefs.

The moment I saw Abdullah's face I sensed trouble. The Commandant at the border post had confiscated his papers and ordered him to report back with us. The problem, a total farce, was a scout's sheath knife which, fool that I am, I had carelessly left on the dashboard shelf where it had been sitting for months, part of the furniture. Of course he simply wanted 'backsheesh', a small bribe, and the knife gave him a face-saving pretext. "With this knife you could kill silently" he declaimed theatrically. "Baksheesh" Abdullah hissed in my ear. "He wants baksheesh". In retrospect and knowing what a pittance he was paid I should have been far more understanding but at the time, I was obstinate and annoyed at being dragged back to the border. "Baksheesh my foot". This response was most unwise because our ship to Piraeus sailed at 14.00 hrs. and it was already eleven o'clock. "You'll miss your boat" almost wept Abdullah and it did indeed seem possible. I said something about contacting the British Ambassador but June saved the day. I can't remember who had the bright idea that Abdullah and I should withdraw but a couple of minutes after leaving her in the Lion's den she emerged with a flower in her buttonhole and the official face was saved. "For the lady I do it, not for you" he spat. A flabbergasted Abdullah implored "how can I become a British citizen" evidently convinced that no Arab could have prevailed. What I did not know, and it would have scared me stiff, was that the Commandant must have 'phoned the docks. I have no idea what he told them but it clearly had the desired effect. After all the other vehicles had been hoisted on board our land rover still sat in lonely state on the dockside. We were politely asked to unload every last item from the jam-packed vehicle that we had so painstakingly stowed for, as we thought, the whole journey. That little operation left everything in chaos; Bedouin rugs, coffee pots with elegant spouts, (one of them a battered old beauty made of copper, blackened with age and doubtless the veteran of countless desert crossings), our two enormous jars of olives, huge copper mansef plate and all our motley gear. Luckily at least we didn't have the wolf cub, jackals and jerboas! The undersides of the mudguards invited finger-tip inspection and they even thoroughly probed the petrol tank. I faced all this with complete composure, even a little humour, because I hadn't hidden anything, but with hindsight my blood runs cold with thoughts of what might have happened had anybody planted drugs or currency, planning to recover them surreptitiously at the docks. After all, the land rover had been out of our hands all the way across Syria.

But at long last they slung it on board in a right old mess. As we sailed out of Beirut bound for Piraeus all we had to worry about was the Russian food, served by female all-in wrestlers, and the forthcoming trek through Greece, the old Yugoslavia and across Europe. Apart from a broken front spring after a magnificently unpaved mountain pass, a minor stoning incident as we passed through a hostile village and 26 savage hair-pin bends on the descent to Kotor we had an uneventful journey. Market-day in Kotor saw the ancient town jam-packed with horse and bullock carts; not a single car to be seen. It could have been 1669 rather than 1969. We approached the town slowly because of all the animals, but not slowly enough for the local policeman who blew his whistle, flagged us down and then, scratching his head for inspiration, demanded to see our passports. With these safely in his eager hands he fined us for speeding at 20 mph. Oh well, I suppose he too earned a pittance and anyway we couldn't argue.

A couple of days later, on a misty morning, we climbed up to Delphi's ancient rectangular Olympic stadium. Beneath the towering crags, with the rising sun shafting eerily through the greyness, and not another soul in sight, we watched crag martins and swifts wheeling and darting. Then Yugoslavia and a ferry across to the lovely little island of Rab where we found a most perfect Naturist beach once, I believe, favoured by Edward VII. Thronged with happy naked brown people, it had no litter, transistors, ice-cream merchants or offensive behaviour – just sea, sand and sun. We tried to camp for the night nearby, in an idyllic old olive grove, but just as the sausages sizzled in the pan a uniformed officer with a kindly face approached and wagged an admonitory finger. "Machen feu verboten". Sadly, we had to up-sticks, sausages uneaten, and move to a proper campsite.

By the time we reached Austria the engine resolutely refused to start or idle all because way back in Jordan a mechanic servicing it had made a pig's ear of adjusting the points. So, whilst June shopped in a picture-postcard village I drove round and round rather than stop the engine. By the tenth circuit people were staring and soon afterwards, faced with push-starting on a German Autobahn, I adjusted the points, which I should have done long ago had I realised the problem.

Then the Channel crossing and after a peaceful night camping in a field outside Dover and a simple run north to our half-ruined Aberdeenshire croft, Azraq joined the lengthening list of loves lost. But other loves waited and the day was yet young.

*Sinai rosefinch*

Palestine sunbird

Yellow-vented bulbul

Arabian warbler

Tristrams grackle

**167**

*Swallows*

*Eiders*

## Chapter 9

# NOSE TO THE GRINDSTONE

*"And now you ask in your heart 'how shall we distinguish that which is good in pleasure from that which is not good?' Go to your fields and gardens and you shall learn that it is the pleasure of the bee to gather honey of the flower. But it is also the pleasure of the flower to yield its honey to the bee"*

'The Prophet' Kahlil Gibran 1972 Heinemann London

It was a dour god who said "Let there be Buchan", cold, windswept and flat, with wide skies and scudding clouds. The sea off Peterhead is pewter-grey and the coldest around Britain. No wonder the warming power of a good malt is regularly appreciated by Buchan's hardy sons.

Quilquox croft, our very first house (discounting the wooden hut on the Bass Rock) sat solidly in a ten-acre field half a mile from the River Ythan which flows into the North Sea over a tidal bar at Newburgh. The nearby dunes are a favoured nesting ground of eider ducks and in May it is impossible to cross them without flushing a nearly-invisible brown duck from her clutch of olive-green eggs, nestling in their bed of soft down. You still can't beat a good eiderdown. The ducklings have a hard time with the predatory herring gulls which swoop repeatedly until the downies are too tired to dive anymore. Then they swallow them whole, headfirst.

Quilquox was a largely forsaken croft that in happier times had been a thriving country store serving the surrounding farms and cottages. Before the coming of the motor car the remoter dwellings depended on journeymen, itinerant joiners, blacksmiths and builders, and on horse-drawn deliveries of groceries, hardware, coal and the like. Quilquox had been a vital cog in the social machinery of this old, austere Buchan. But times change and in 1966 the widowed Mrs. Milne wanted to move into Aberdeen and advertised the croft for sale in the 'Press and Journal', on which James Naughtie cut his formidable teeth. Luckily for us it was just before offshore oil in the North Sea turned Aberdeen upside down, destroying the old order and pushing prices sky-high. From a farming and fishing town little changed over the centuries, to fancy restaurants, boutiques and strip-joints. New buildings sprouted everywhere, utterly changing the skyline. The Aberdeen airport that we first knew suddenly seemed quaint and inadequate for the booming oil-trade and the city built a new one. Road traffic increased mightily and Aberdonians had to get used to crawling in and out of the city. In a few short years the staid silver-granite capital of north-east Scotland changed utterly. Welcome to the 1970's. And now, the 'seventies seem old-fashioned; a crazy pace of change.

*Quilcox Croft, Ythan Bank, Aberdeenshire (1969) our first house*

*And as a shop and horse drawn delivery in the 1930s*

We bought Mrs. Milne's 'butt-and-ben' with granite outbuildings and 10 acres of good grazing in January 1967 and straightaway set sail for Christmas Island and Abbott's booby. It was all completely mad. Barely two minutes before we pulled out of Aberdeen railway station on the night sleeper to London we still hadn't bought it. Having that very day finalised the details our solicitor rushed onto the platform clutching the deeds for us to sign. It was a serious challenge to find a couple of passengers trusting enough to act as witnesses when we signed the deeds and nowadays, with scams everywhere, you wouldn't stand a chance. God knows what the signatories thought they might be letting themselves in for, but our staid Aberdonian solicitor saved the day. That done, we settled back, the proud new owners of a wee patch of Buchan. Unfortunately we returned nine months later penniless, jobless and with major house-building to do. But all that was still well below our horizon.

November in Buchan paints everything in shades of bitter grey. Vast skies arch over flat farmland broken by pylons, telegraph poles, moaning wires and miserably wind-blasted copses of fir, sycamore and beech. Our blind, empty and ravaged croft merged seamlessly into a nettle-topped mountain of old bottles and rusty cans, the detritus of decades. This heap leant against a decrepit old railway guard's van which had been used for storing paraffin. Like an idiot I put a match to it and almost razed the steading. I knew granite wouldn't burn but I forgot about the old grass in the gutter. The neglected garden nurtured masses of ineradicable coltsfoot and ground elder, surely two of the most obdurate weeds known to woman, far worse than nettles or docks, although we were blessed with those, too, amongst the roots of ancient, lichen-encrusted blackcurrant bushes. All this and a disembowelled cottage, for Mrs. Milne, entirely properly, had 'rouped' (Scottish term for auctioning) all the movable contents, which included the shop counter, brought us up with a jolt. We sat on old packing cases, our only furniture, and thought about it. The Buchan sleet mocked our still-fresh memories of the tropical Galapagos, our delightful months on uninhabited Tower Island, and the vibrant jungle of Christmas Island we had so recently left behind. What now, you feckless fellow? We were in this pickle because of a chap by the name of Wilson, Harold to his friends. The promised job at Aberdeen University had evaporated when his government cut university funding and with it my proposed lectureship. But thanks to the redoubtable Max Nicholson rescue was at hand and by the time we returned from the Jordanian desert eighteen months later things had looked up. So here we were in Buchan again – the land of the corn bunting, and this time with a job.

*Corn bunting*

You probably couldn't do it nowadays but in 1969 dear old Prof. Wynne-Edwards, courteous and softly spoken but gimlet-eyed and very much an alpha-male, appointed me to a zoology lectureship 'in absentia', whilst we were still in Jordan. My bread and butter was to teach basic zoology to first-year medical students, many of whom entered the Faculty of Medicine knowing nothing about zoology and caring less. They had no idea what an animal cell was or how a heart or kidney works, whilst systematic zoology, evolution, genetics and animal physiology were closed books which they had no desire to open. Maybe they thought they would move straight into heart surgery, or, at least, skull trepanning, which as our Prof. Ritchie liked to say, is a bit hum-drum, hell on the feet and therefore suited to younger practitioners.

For my first few lectures the class of 140 aspirant doctors eyed me with palpable hostility. I was not even a medic and I sure as hell wasn't telling them anything they thought remotely relevant to their potentially glittering careers. But eventually we managed to establish a modest rapport and by the time we touched on animal and human behaviour, my personal interests, a few of them became quite keen. It was a steep learning curve for me, too. This was my first shot at conventional teaching, far different, and harder, than one-off lectures on one's own specialities. I had uncomfortable memories of my own undergraduate days at St. Andrews when first-year medics attended some of our zoology classes and gave the poor old lecturer, admittedly excruciatingly dull, a hell of a time with slow handclaps, foot-stamping and audibly rude comments. I could do without that.

It was much more fun to organise the teaching of animal behaviour to third-year zoology students which I did jointly with Ian Patterson. He, too, had been one of Niko Tinbergen's students at Oxford in the very early 'sixties. Clear-thinking, concise, practical and modern in outlook Ian nicely complemented my romantic and overly-ambitious ideas for undergraduate behaviour projects. The students liked our concentrated six-week course of lectures and practicals, especially the opportunity to choose a small research project on which they wrote a mini-thesis, or in some cases not so mini! We designed about thirty projects intended to test their ingenuity, powers of observation and ability to analyse the results. I had the chore of assessing them all and with sixty or more students in the group it took some ploughing through. It was their first opportunity to do their own thing and they loved it. They tackled a wide-range of projects, both experimental and observational, using fish, rats, birds, sheep, cattle and humans. The latter involved covert observation of, for example, the ways in which we greet one another, the differences between males and females in the way they carry books (females tend to clasp them across the front of the body; males more often use a pinch-grip), the size and sex-ratios (male/male, female/female or male/female) of groups of children leaving a school at lunch-time; the displacement activities such as head-scratching, finger-tapping or yawning by drivers held up at traffic-lights and much else. Some students designed their own projects. One highland laddie opted to study red-deer which meant a daunting amount of hill climbing and stalking. They found the challenge hugely stimulating, but alas the intellectual climate was changing. The safety-first culture and the fear of possible compensation claims reared their ugly heads and even after several successful years and no fatalities the university administrators began to fret. An Edinburgh university student, perhaps further removed from his arboreal forebears than ours, fell out of a tree and alarm bells began to ring. So Ian and I were asked to restrict our projects to the laboratory; a pity, but in this stupidly litigious age perhaps inevitable. What if a pigeon-watcher on Union Street fell under a bus?

One of the best things about a University job is that you can, or could, pursue your own research at your own pace, free from commercial or any other pressure. Long-term fieldwork like mine simply cannot produce quick results but there is no other way of answering fundamental questions to do with a seabird's productivity, lifespan, long-term pair-relationships and population dynamics. So for me it was back to the dreich old Bass to pick up the gannet threads. Aberdeen University zoology department, benign and supportive under Mike Begg (the Reader, eventually to be my post) and Prof. Wynne-Edwards, who was often away in London in his capacity as chairman of NERC, provided me with an Avon inflatable so that in fine weather I could paddle around the base of the Bass Rock counting gannets. As described in Chapter 2, a great cavern, accessible only by sea, runs beneath the rock from west to east. In its damp and gloomy recesses the French sympathisers of Catholic James II hid provisions for the small band of Jacobites who had cunningly taken over the fortress in the 1690s and held out for three long years. This secret cavern opens at the base of the east cliffs where its mouth is bisected by a rocky outcrop, in summer clamorous with gargling guillemots that whirr out on frantic wings, plop heavily into the sea and fizz down into the depths as pale green blobs. Partway through the passage a small pebble beach has been thrown up by the surge. Grey seals sometimes drop a pup there.

One of the third-year zoology students at Aberdeen in 1973, Sarah Wanless, a tall athletic girl from Scarborough, had been smitten by gannets. She asked if I would supervise her thesis at Aberdeen as part of the requirements for an Honours Degree. She wanted to do a small study of the little gannetry at Bempton, not far from Scarborough. Bempton's cliffs are famous from the old days of the 'climmers' who descended for seabird eggs. The Bempton gannets, at that time a mere handful, probably come in the main, if not entirely, from the Bass not far to the north. Bass birds regularly fly past Bempton on fishing trips and on migration and one of our colour-ringed youngsters from the Bass eventually set up home at Bempton. Shortly after her finals NERC as it then was awarded me a grant to finance a Ph.D student to work on the gannets of Ailsa Craig. The nub of my grant application had been to compare their breeding ecology with that of the Bass gannets to see if there were significant differences between the east and west coast. Both colonies, famous, long-established and at about the same latitude, have grown enormously in the last half century. In 1905 each of them held a mere 6,500 'pairs'. Now they are many times that number.

Sarah was my obvious choice for the Ailsa work. It was a tough assignment but she was a tough

*Sarah Wanless and her dog*

lass. I don't know how many times she slogged up the long, steep track past the old castle to the cliff top but it must have been hundreds. To add to the difficulties her gannets were far warier and less approachable than the Bass birds. Physically, the Bass had been a cake-walk compared with Ailsa. But fortunately she had a proper house to live in, for Jimmy Girvan who used to quarry granite to make the famous green-veined curling stones, had retired to Girvan and he generously allowed Sarah to use his old cottage. Many times, after a hard slog up the rock on a cold, grey and wet day, she must have blessed that cottage with its cheery fire of driftwood.

In 1960 I chose the Bass partly because it is much more accessible than Ailsa. As it turned out this easier access and the visitors it encouraged made the gannets much bolder than the paranoid Ailsa birds, a tremendous boon. You can imagine the hassle involved when the birds are so wary that they flee before you get anywhere near, to say nothing of the effect of disturbance on the birds themselves. At worst they may lose eggs or chicks and even if they don't, their breeding regime is greatly affected.

As it happened, Ailsa rather than the Bass had been my seminal introduction to gannets. Courtesy of the Girvan trawler 'Selina II' a friend and I first set foot on Ailsa in July 1953 armed with 200 gannet rings, a kitbag full of pan-loaves and a dozen tins of pemmican meat extract. At that time Jimmy Girvan was still quarrying the beautiful granite for curling stones – work which at one time employed up to 60 men. At the top of Ailsa's slipway stood a large old shed, now long gone, in which Mrs. Girvan served tea and delicious scones to day visitors. Her goats grazed the short, sweet turf on the foreshore and at low tide browsed on seaweed. Ringing gannet chicks on Ailsa's rotten ledges was the formative experience that shaped much of my life.

As Sarah settled in and began to explore the base of Ailsa's cliffs she found that she had a gruesome task on her hands. At the base of the cliffs amongst the boulders and vegetation squatted adult gannets which would never fly again. They faced her defiantly enough, as a gannet always will, but they trailed a shattered wing. They had crashed after take-off from the high cliffs though it was not clear why. It may have been due to the shape of Ailsa and the extensive rocky foreshore. By contrast the Bass cliffs drop sheer into the sea which, in a crash-landing is kindlier than boulders. Perhaps the unfortunate Ailsa birds ran out of airspace. Gannets are heavy, with long, narrow wings and they cannot float like a gull. Indeed one good reason why they prefer to nest on the cliffs of offshore islands is precisely because of the helpful wind which usually blows. Whatever the cause of the accidents Sarah had to kill literally hundreds of injured gannets which, otherwise, would simply have starved to death, and gannets take an age to starve. They are adapted to survive for weeks in winter when stormy weather may make fishing impossible. And, perhaps even more than most large birds, gannets are horribly difficult to kill. Chicks fell too and were eaten, some of them whilst still alive, by the hordes of rats which at that time infested Ailsa. It is only comparatively recently that Bernie Zonfrillo eliminated them. He did it by skilful and persistent use of poisoned bait; a huge task, for if even a few survive you have wasted your time. As a direct result of his efforts the puffins, which used to nest in thousands before rats got ashore, are now returning.

So, with the Bass work still going, Ailsa chiming in and with a full load of teaching, what time was left for our dismal little croft during those early years at Aberdeen? In brief, we were simply idiots. Long accustomed to doing things for ourselves and with an old-fashioned aversion to debt, which, alas, our children have not inherited, we saved ludicrously small sums whenever we could – a few miserable pounds by digging out the concrete shop floor, barrowing it across sloping planks and tipping it outside the kitchen where it lurked in an unsightly heap for months. This was mid-winter in Siberian Aberdeen. Two massive holes had been knocked through the outer wall of the putative living room to make way for windows and the snell Buchan wind drove sleet directly into our morgue-cold house, brusquely breaching the pathetic drape of tarpaulin with which Tawse brothers, the local builders, had covered the gaping holes to mollycoddle us effete southerners. We had no electricity or gas, no fireplace and were scrabbling around on the bare earth, little better than cave dwellers. In fact Neanderthal caves were undoubtedly much cosier. Later, when the transformation was more or less complete, a local, eyeing the result wistfully, remarked sourly, "aye, it must be fine to hae monny".

In a stroke of genius, we rescued five blocks of beautiful, rough-hewn, silver granite from an Aberdeen City municipal tip. One of them had been the keystone in an arch, perhaps a church window, and for the cost of hiring a lorry with lifting gear the granite was ours. With silver granite cobbles for a hearth that arch made a magnificent fireplace. On a frosty night you can't beat a deep peat fire glowing redly and fragrantly. We could have roasted an ox in that fireplace and I don't know why we didn't at least try a sheep or a pig. But we did have a ceilidh in the living room, with a real fiddler. Some nifty traffic diversion round the foot of the open staircase gave just enough space for two cramped eightsome reels.

These days it would cost the earth to build a tennis hard court and we possessed hardly a molehill, but our croft sat in the middle of a ten-acre field. It was on a bit of a slope but we could dig out a level plot for a tennis court. We bought two posts and a second-hand net for ten pounds and hired a man from Aberdeen to dig out the court, lay field drains across it and lay down 300 tons of hard-core. The main challenge was the actual playing surface. En tout cas was impossibly expensive and grass totally impracticable so we settled for fifty tons of silver granite dust from Craigenlaw quarry in Aberdeen, waste from cutting and polishing headstones, granite cobbles and the like. Fifty tons made an impressive heap even if it doesn't sound a lot. The main cost was the haulage. Our competent contractor then toiled back and forth with his ancient little dumper, transferring granite dust from the 50 ton heap to our embryonic court where he spread and levelled it before hiring a mechanical thumper to compact it. Hey-presto! Once we had laid down the plastic lines we had a tennis court. It wasn't bad to play on once it had been watered and rolled but if it dried out the surface crusted-over and the ball broke through and buried itself. So in dry weather we simply had to rake, water and roll. Our tennis friends were never on hand for this.

There remained the thorny problem of fencing. But one fine day a friendly and entreprenurial neighbour rolled up with a lorry load of fifteen foot larch poles which had somehow cadged a lift. This left only the boundary netting for which we used rot-proof garden stuff. That home-made court gave us enormous pleasure. At the top end, as a consequence of digging out the level playing area in a sloping field, there was a steep bank in one corner of which we excavated a snug recess, ideal for barbecues and sun-bathing. More than one Buchan farmer, and with good reason, looked askance at this wicked misuse of good grazing land which, over the many decades, had fattened thousands of bullocks. And by Sassenachs too. It was only a small bit of our field but it didn't improve their opinion of 'southerners' which, to us Yorkshire folk, is an insult.

Just across the River Ythan from us, lay Haddo House, now the property of The National Trust for Scotland, but at the time owned by Lord Aberdeen. Once a Buchan bog, the Haddo estate is now richly varied woodland and farms. Haddo's large wooden concert hall was the venue for many plays and concerts. June Gordon (Lady Aberdeen) conducted the Haddo orchestra and gracefully roped in Richard Baker to help with productions. The annual Haddo Ball was a fixture early in the New Year, a typically rural Aberdeenshire jamboree with a substantial supper. A fair number of the male combatants felt honour-bound to drink the bar dry but the behaviour of the country-folk, many of them farmers, always remained impeccable. Traditionally the evening ended with a maudlin rendering of 'Auld lang syne' before everybody tumbled out into the frost and snow to wend their erratic ways deep into the icy countryside. Haddo was by no means the only Country Ball. The Ladies of Ellon, Mintlaw, Methlick and several other such august bodies all held their Annual Ball, a tradition from the old days when there were none of the modern diversions to lighten the long, dark, bitterly cold and snowy Buchan winters. People made their own amusements which usually involved a dram or two.

Those Aberdeenshire years were the busiest of my life. We renovated the old croft, fought a bitter war against weeds and rabbits, raised our twins through their early years whilst I tried to earn my modest university salary. I made several trips to Christmas Island and, above all, wrote scientific papers and a massive tome of nearly a thousand pages about gannets and boobies. Mostly based on my own work it also involved a lot of scientific literature. It was long before the day of the computer and I wrote everything in longhand, three or four

*Boxing hares*

times, before typing it and preparing hundreds of figures and tables. The long-established and prestigious publishers, Oliver and Boyd of Edinburgh, set it up in 25 tons of metal letter-press but, understandably fed up by my failure to deliver on time, they ditched it. Luckily for me, Aberdeen University agreed to take it up and they published it via Oxford University Press. I was very, very lucky. If they hadn't taken it on I'm sure nobody else would. At the time it seemed as though my life's work was at stake. Fortunately I didn't give a fig about the money since Aberdeen university took all the Royalties. Best of all, dear old Tom Jenkins of Oliver and Boyd, the ultimate perfectionist and very much of the Old School, acted as editor and midwife, paid by OUP. I imagine there cannot be his precise equivalent in the whole of British publishing. He mocked-up the entire 1,000 page book on blank pages, in properly bound volumes in which he positioned the text, represented by simple lines, and the position of each one of the hundreds of line-drawings, graphs, tables and photographs, each in their appropriately reduced size. It was a gigantic effort and a tribute to his half-century's experience with Oliver and Boyd. He arranged for two gold-leaf copies to be specially bound in blue calf-leather with embossed gannets and boobies in colour, one for him and one for me. I sometimes wonder whether the comparatively few times books like this are referred to justifies all the effort.

A large chunk of us went into the renovation of Quilquox croft and it was a wrench to part with it. Nowhere else was going to be quite the same. We had bought it as a dilapidated 'butt-and-ben' with decaying outbuildings, Scotland's finest heap of vintage tins and bottles and the sturdiest nettles for miles around and turned it into a home full of character. It saw the children through their formative years and in it we navigated some choppy marital waters, all the more painful to me after so many idyllic and adventurous years. But it was all part of the rich tapestry, or so they say. And there were adventures still to come.

*Red deer jousting*

*Red-footed boobies roosting*

Chapter 10

# ALDABRA ATOLL

*"From their slowly sinking foundations of ancient volcanic mountains, the creatures of the coral shoals have erected the greatest organic structures that exist. Even the smallest atoll far surpasses any of man's greatest building feats, and a large atoll structure in actual mass approaches the total of all man's building that now exists"*

John D Isaacs 1969 In 'The Ocean' Scientific American

My British Ocean Territory Administration ticket No.420, dated January 25th 1973, was actually issued in 1974, the year in which our twins were born, and valid for voyage No. 78 on MV 'Nordvaer', from Mahe in the Seychelles to Aldabra, a speck of sea-fretted coral in the Western Indian Ocean. Aldabra, the great seabird metropolis on which tens of thousands of frigatebirds converge to build their flimsy nests in the mangrove swamps. The air is seldom free of these great black pirates, forked tails scissoring as they deftly control the shifting air currents. It was the frigatebirds that drew me.

Early on a bitterly cold January day I packed my bags miserably and prepared to abandon my wife and newborn twins to the dreich depths of winter in Aberdeenshire Buchan, grey and wind-swept on that Siberian day. My bags for Aldabra squatted on the striped Bedouin rugs, trophies from the Jordanian desert. The smell of goats and camels still lingered, redolent of the Beni-Sakr's great black tents in which we had sipped tiny cups of bitter green coffee whilst sitting uncomfortably cross-legged for longer than the average Briton can cope.

This venture was all about frigatebirds. The prestigious Royal Society of London had established a Research Station on Aldabra at a time when that atoll's magnificent wildlife was threatened by plans, mercifully never fulfilled, to build a military base there. As so often, American interests were involved. It wanted the base. Wildlife, however rare and wonderful, even if found nowhere else on earth, has no chance against the military, or even, sometimes, against commerce. But by becoming involved in Aldabra the Royal Society at least showed its concern. Aldabra is one of the world's major frigatebird colonies and my job was to suggest further lines of research after Tony Diamond's fruitful initial work.

Because they are so awkward to study, frigatebirds were still mysterious even though spectacular and comparatively common. They nest in mangrove swamps, which are difficult places to work in; they breed incredibly slowly, taking well over a year to rear one chick; they tend to lose many of their eggs and chicks; and just to complicate matters they change nest-site and mate at each breeding attempt. And to cap all, frigatebirds are devilishly difficult to mark. Without recognisable individuals you get nowhere. Their tiny legs are more or less hidden in their belly feathers, so rings are useless. Back in the 1950s, as part of the British Ornithologists Union's Centenary Expedition, Bernard and Sally Stonehouse had tackled the frigatebirds on Ascension Island in the South Atlantic but they had barely scratched the surface. Much the same could be said about our own work in the Galapagos in 1964 although we had dispelled some mistaken ideas. The main snag had been our inability to recognise individuals. Was that male the same bird we had seen displaying there yesterday, the day before that, or last week? We simply could not tell. Was the bird now sitting so placidly on its nest the rightful owner, or an intruder who had thrown out the egg and usurped the nest? Again, we didn't know, although we did know such conflicts occurred. Why was that male attacking and killing a defenceless chick? Don't ask me; no idea. Add to all this, and many other questions, the sheer impossibility of even reaching nests high in the mangroves and in addition the huge loss of eggs and chicks that would surely result from human intrusion and it was obvious that the simplest tasks would be fraught with difficulty. Out of the window goes reliable figures for breeding success, that basic component of ecological studies. Unless you know how successfully they breed it is impossible to make any sense of a long-term population study, and conservation needs this sort of information. So, to come to the point of my visit, it was crucial to find out if it was practicable to mount a decent study of Aldabra's frigatebirds. Maybe it wasn't.

*Frigatebird chick*

Nobody could have worked on uninhabited Aldabra without the logistical and other support of The Royal Society and its base on the island. I was lucky to have that support, and whilst my colleagues in Aberdeen would be splashing through the grey slush, to the soulless concrete block that was the new, costly, Soviet-style Zoology Department, there to hold forth on insect-hormones, echinoderm tube-feet, tapeworms, marine life and molecular biology (an up-and-coming discipline in those days which has since up and come), I would be speeding over blue lagoons towards the green mangroves and the frigatebirds. But I was sunk in gloom and foreboding. Often in the weeks that followed my thoughts flew back to an Aberdeenshire croft, the cold grey cobbles outside the stable, the keening of the wind in the wires and the laborious humping of coal and peat into the house, which should have been my job. And to a wife with new-born twins. This was the first of my expedi-

*The young frigate-bird is highly vulnerable in the low shrubs on Tower Island, despite vociferous protest*

tions she had missed, and it felt all wrong, and not only to me, for she would have loved it.

The trip began badly on a leaden January afternoon. In 1974 Aberdeen airport was still little more than a shack. The oil-industry, though beginning to boom, had yet to foul-up this ancient silver-granite city. But before Andrew, my farmer friend, had set off back to feed his pigs the first snowflakes began to settle innocently on the runway. Long before the swine were browsing in the trough the runway was well-covered and a North Sea blizzard was brewing in the rapidly darkening winter afternoon. The flight to Heathrow was delayed but with three hours in hand for the Seychelles connection there seemed no need to panic. Two hours later, and still grounded, panic seemed quite in order. A telephone call confirmed that the Seychelles flight was due to leave on time and the point at which it must inevitably do so without the frigatebird man came and went. With unbelievable perversity the flight from Aberdeen promptly took off without me and the Seychelles flight was delayed for so long that I could easily have caught it. Had I but known it, perversity was the motto of things to come. "No man, having put his hand to the plough……" But turn back I did, back to an astonished wife, a new-born twin on each arm.

Where was the original Garden of Eden? Before the jumbo jets rumbled and grumbled down Mahe's runway, so close to the sea that one wing literally hangs over it, the Seychelles must have been a front-runner. But mass tourism makes a spectacular mess of simple societies, especially on small islands. If I hadn't already realised this when the taxi hordes besieged Mahe airport, the first hour with my irascible landlord soon convinced me. I hadn't been in his stifling kitchen for five minutes before he unloosed a thoroughly justified tirade against tourism and Seychelle's politicians. The old peasant-based economy of the islands had been shot to pieces. Absurdly, and almost beyond belief, vegetables which formerly had been produced in abundance and sold, split-fresh and dirt-cheap, were now being flown in from France and sold, stale, at prices which only those islanders with jobs related to tourism could possibly afford. Meanwhile hundreds of productive plots lay empty and overgrown whilst their owners drove taxis or worked in hotels. It was sheer madness and it must be the same in many parts of the tourist world. To my chagrin my irate landlord included me, as a presumed tourist, in the whole sorry mess though few real tourists would have frequented his cockroach infested, cement-floored cubicles, or tolerated the relentless fervour of the steel band which drove the angels of sleep far from me until dawn, even if the humid heat had not done so. Food, there was none. Clearly, from his clientele, (me, a globe-trotting hippie and some ladies of easy virtue) his establishment was no part of the brave new order. Still, he was a fountain of knowledge. From him I learned that tuberculosis, amoebic dysentery and gonorrhea were the commonest diseases on the island. Probably it was the British, true to form, who imported the latter, along with tea-at-four and portraits of the Royal Family. Or maybe it was the French. Anyway British influence and the English language dominated Mahe, though the French contributed the names of the islands (Praslin, Silhouette, Aride etc.).

Somewhere, waiting for me, lay mysterious Aldabra, home of flightless rails, giant tortoises, Seychelle fodies and a whole kingdom of seabirds. But before the boat sailed I had the chance to visit tiny Cousin Island, an ICBP Reserve, and at that time wardened by Tony Diamond, another seabird enthusiast who, as I said earlier, had worked on Aldabra's frigatebirds. I had last seen Tony in Marischal Quad, Aberdeen, a picture of misery hunched against the penetrating winter sleet, his Seychelle's tan paling rapidly. Now, bronzed, bare-footed 7and manning an inflatable, he sped towards us across the deep blue channel separating Cousin from nearby Praslin. Life on a tropical island can seem blissful. Away with imprisoning walls; outdoors and indoors can merge instead of the hostile outdoors shut off from a cosseted indoor world. There were roomy verandas, shutters thrown wide, spacious open-plan interiors and lazy warmth. The riot of colour and greenery and the sparkle of blue sea invaded the house. Little wonder many ex-pats find, often sadly, that a return to the dreary greyness of a typical British winter and the tedious round of pub, supermarket and telly don't add up. They feel not to belong anywhere.

*Baby turtles*

But to Cousin Island. Towards dusk, wandering happily along a small sandy beach, I stumbled across a magnificent hawksbill turtle digging a hole in which to lay its eggs. She was so engrossed that I crawled to within inches without disturbing her. The digging movements were completely machine-like. Down went her hind flipper, followed by a scooping movement, a twist, and up came a small heap of sand to be put aside. The rhythm was hypnotic. The excavated sand lay in the hollow of her flipper as though in a sensitive hand which she withdrew so delicately. The finished hole, roughly bottle-shaped with compacted walls then received maybe a hundred or more white, soft-shelled eggs before the turtle, weeping copiously, covered them with sand and crawled wearily back to the safety of the sea. Her enormous tracks were soon wiped out by the incoming tide.

Behaviour like the turtle's hole-digging is instinctive. And as the famous Austrian ethologist, Konrad Lorenz showed long ago, animals are strongly inclined, one could say almost impelled, to perform these programmed actions, so much so that if the appropriate stimulus does not come along they will eventually perform them in response to the tiniest cue, or even none at all – 'in vacuo' as he said. He cites a captive bee-eater that, denied any opportunity for dealing with a bee, eventually went through the entire process of de-stinging an imaginary one, performing the appropriate wiping movements of its bill against its perch. Having once started digging, my turtle would not have stopped even if its hole had been miraculously completed in a couple of scoops. Two scoops would not have dissipated its urge to dig and the turtle would have continued regardless until something like the normal quota had been performed.

But it was not the stolid turtle that most excited me on Cousin Island, it was the angelic fairy tern, as dainty as its name. Cousin Island evidently suited it whereas it does not grace the Galapagos Islands and is equally disapproving of Christmas Island, though I could not see why; there were innumerable nesting places and the island suited brown noddies (also terns) well enough. Fairy terns scorn the very notion of a nest. Such cavalier disregard may be all very well for birds which lay their eggs on mother earth or even on a cliff ledge, but it is rather trickier when the nest site is a bare branch or a fork in a twig, which is what fairy terns make do with. Only about a quarter of all eggs survive long enough to hatch, and even this modest success seems miraculous. In some localities fewer than one egg in ten produces a free-flying youngster. Such low productivity is common among tropical seabirds, few if any of which habitually fledge young from as many as half the eggs laid, whereas many northern and Antarctic seabirds fledge 60-90%. For such birds a breeding success of 10% would be calamitous. Tropical seabirds often fail simply because their food suddenly disappears whilst they are in the thick of breeding. This rarely happens in temperate waters. On top of this, and in addition to its death-defying nest sites, fairy terns on Cousin lose eggs to skinks and to Seychelle fodies. Mainly though the terns lose eggs, and even large chicks, because marauding adults of their own species attack them. Nor is this totally perplexing behaviour unique to the terns. I often saw it in frigatebirds and masked boobies and occasionally in gannets and suspect that it is widespread in seabirds and perhaps even more widely than that. For instance, it occurs in swallows. An obvious possibility is that it releases the female concerned back into the pool of available mates, but there are snags to that interpretation. Originally it was the havoc wrought by apparently 'spare' frigatebirds on Tower Island in the Galapagos that intrigued me, but if there are unmated and physically fit birds of both sexes available, what is the advantage of sabotaging somebody else? Even if the attacking male could then mate with the female who had been so summarily relieved of her domestic duties, which is unlikely, he would only get a partner who had already spent a lot of time and energy and would therefore be less 'fit'. But the fact is that 'spare' males do enter breeding colonies and do disrupt breeding pairs. I doubt if there is any sizeable seabird colony that does not include some non-breeders, including transient, unpaired birds, mainly males, and it is probably these birds that do the damage.

Like many other terns and gulls, fairy terns often 'kiss-preen'. That this apparently affectionate behaviour involves aggression is betrayed by the rapid stabbing action of the male. In the gannet, 'kiss-preening' can be extremely rough, often grading into actual attack by the male when the pair-relationship is fraught and on a knife-edge, as it often is when the partners are new to each other. In humans, too, aggression is overwhelmingly from male to female. For no good bio-logical reason, most of us seem to believe, and no doubt want to, that humans lie quite outside the norms and principles that govern the be-haviour of other animals. To most biologists who know a bit about evolution and ani-mal behaviour and are not indoctrinated by religion or politics that seems a strange idea. But crazy notions may be part of being human!

*Fairy tern*

Seabirds have many enemies, including other birds, reptiles, crabs, cats, dogs, pigs and rats, but on Cousin noddies and fairy terns fall victim to a plant, the murderous Pisonia tree in which they nest. Hundreds of its spiky seeds settle onto the birds and bind their feathers together until, in their struggles, they fall to the ground and be-come further entangled, hopeless bundles and easy prey for the land crabs. Pisonia trees and terns must have co-existed for donkey's years but it seems that the birds still have not learned to avoid them despite the intense selection pressure which must have operated. This is quite different from the usual predator/prey relationship in which the predator progressively evolves better catching behaviour to out-wit the si-multaneously evolving escape behaviour of its prey. There could not be any selection pressure favouring Pisonia trees that kill terns! But selection should certainly have strongly favoured terns that removed seeds the moment they landed on them, for it is only after several seeds have meshed with the barbs of the feathers that the bird is in danger. Yet this simple seed-removing behaviour has not evolved. There do exist un-usual situations in which birds have managed to come up with appropriate counter-actions. For instance, the chicks of the black-footed albatross are sometimes in danger of being buried alive by wind-blown sand, so as soon as it begins to pile up against them they vigorously kick backwards to get rid of it.. Without this reaction they would be buried alive and indeed this sometimes does happen to chicks of the Laysan alba-tross, which, even when nesting on the same island as the black-foot, lack this sand-kicking action. But as for the terns, generation after generation continue to nest in trees which kill them. Odd, and extremely distressing.

This beguiling interlude on Cousin Island soon ended and M.V. 'Nordvaer' began to make serious departure noises. Its timetable had to be fairly loose because, like all general cargo boats plying between small islands it served many interests, tarrying as required here and there. It reminded me nostalgically of the old 'Cristobal Carrier' which used to weave its slow and erratic course between Guayaquil, the hell-hole of the Pacific, and San Cristobal in the enchanted Galapagos Isles. But whereas I feel affection for that asthmatic old landing barge I abhor 'Nordvaer'. However, in January 1974 I had no inkling of what that abominable old rust bucket had in store for us. She had been freshly decorated below decks and the overpowering smell of paint exacerbated the slight nausea brought on by her incessant rolling. A large pig had the run of the port side; whether it was for delivery or for fresh bacon on board I wasn't there long enough to discover. On the second day out of Mahe a large, oily swell under a black sky ominously tinged with green coincided with the dread word 'monsoon', clearly audible above the static in the radio operator's cubby hole. Luckily, it never came to anything.

Apart from a few masked boobies, wedge-tailed shearwaters and royal terns the sea seemed empty. Seabirds, even from huge colonies, easily disappear in the wide expanses of ocean. On the other hand they do congregate in large mixed flocks in the productive areas, leaving vast expanses of sterile 'blue water' empty. So perhaps, despite appearances to the contrary, there may be some competition for food. Often, though, the indisputable signs of food shortage which sometimes are evident in colonies - chicks starving, nests and eggs abandoned – do not mean that the food in the surrounding sea has been depleted by seabirds. Often, the shortage is merely temporary, a marine phenomenon quite independent of the birds, and lasts for only a few weeks before things suddenly return to normal.

'Nordvaer' had to collect copra from Assumption Island and put stores and labourers ashore. The heavy surf creamed and crashed on the shingle and it took 20 or more sturdy islanders to hold the bows of the longboat facing the breakers. I was more than a little interested in Assumption Island. It had long been high on my 'wanted' list for it was on this island that, in 1892, the American naturalist W.L.    Abbott is reputed to have collected the first ever specimen of Abbott's booby, the bird which had played such a central role in my life. The collection site 'Assumption Island' is written plainly on the label. Nowadays its sole nesting place in the world lies thousands of miles to the east, on jungle-clad Christmas Island but fossil and historical evidence shows that at one time it did indeed nest in the Western Indian Ocean, on Rodriguez and Mauritius. But, and it is a big 'but', it emphatically did not nest on   Assumption, whatever the ticket says! How can I possibly know? Well, on Christmas  Island it nests high in jungle canopy but on Assumption there is, and was in W.L.     Abbott's time too, only scrub on low dunes, totally unsuitable for this exceptionally long-winged booby which often needs considerable free-fall before it becomes air-borne. Assumption Island could not have provided that essential. So why does the old label on this first-ever specimen say 'Assumption Island'? The answer seems simple. The labels on the specimens became crossed, which can easily happen on a major collecting expedition especially when, as in this case, the specimens were not properly looked at and classified until years later, when memory was cold. It has to be remembered that Abbott collected maybe hundreds of birds from several islands and circumstantial evidence suggests that when he later wrote of Abbott's booby "--- a few nest there" he was actually referring to Glorioso, where he did indeed collect birds and which at that time was covered in high trees. The French later felled them all. The absolute clincher, for me, is that Abbott described the voice of the booby he collected on Glorioso as "bull-like", which is exactly what Abbott's sounds like, and no other booby does. So my vote goes to Glorioso and nothing I saw on Assumption Island persuaded me otherwise. Whether the ornithological establishment will embrace this historical correction, even from the person who knows Abbott's booby better than anybody, is perhaps another matter.

*Assumption islanders' holding the dory bows-on to heavy surf*

Long before Assumption and Aldabra atolls could be seen they betrayed their positions by pale green reflections in the sky, a fact doubtless used by native navigators like the Polynesians who crossed vast tracts of Pacific in their large out-rigger canoes. Aldabra opened its ornithological account in familiar fashion with the agonised "aa-ark" of a red-footed booby fleeing frantically but hopelessly before the effortless pursuit of a 'man-o-war' bird. This tinny and prolonged screech, as grating as the hoarse "yakking" of the piratical frigatebird, had been the despairing background noise to our daily life on Tower Island in the Galapagos. Sometimes, beyond the beach in the sun-silvered scrub, we found the dried corpse of an unlucky booby driven to a desperate crash-landing and ending up inextricably entangled in the scrubby branches of a Palo santo tree. Why do so many waders possess melodious calls whereas sea-birds are as raucous as pop-singers? It may be because waders use their voices in territorial song-flights and, at the amplitude which a small bird can produce, a clear whistling call carries further than a low-pitched note.

At the red-foot's despairing call, and the memories of Tower Island and June, my thoughts flew to a cold and draughty Aberdeenshire croft where, at that very moment, rooks were cawing their way to a winter roost in a local copse and she was coping with month-old twins. Something else was worrying me too! It may seem a mere detail, but I had omitted to obtain the consent of the University Court for my little trip in term-time into the back of beyond. Now that I was on the verge of the unknown, the return date seemed ominously less dependable than before. As things turned out, it certainly was. Fortunately for me, Prof. Wynne-Edwards, with typical kindness, rescued me again. Knowing how careless I was in such matters he had informed the University Court on my behalf and had supported my part in the expedition. University lecturers may beef about low pay but they do (or did) enjoy many advantages over their commercial or civil service equivalents. However, things seem to have changed since 1974 and perhaps universities are no longer what they were. It may seem incomprehensible to to-day's average employee, but throughout my university career I never received a direct order. Maybe a request, or a suggestion, but not an order. Of course this civilised relationship made most of us pull our weight.

Aldabra is a somewhat pear-shaped atoll, 34 by 14.5 km. The rim of limestone islets encircling the lagoon is heavily dissected and covered with low scrub and a few tall Casuarina trees. Inside the lagoon there are extensive mangroves. North Island is inhabited by giant tortoises —"land turtles of so huge a bigness which men will think incredible, of which our company had

*Giant tortoise*

small luste to eat of, being such huge, deformed creatures and footed with five claws lyke a bear". Whalers started operating around the Seychelles after 1823 and in 1842 two ships from Hamburg collected 1,200 tortoises, some of which weighed more than 400 kg. Charles Darwin, who was familiar with giant tortoises in the Galapagos, championed their conservation and they are still numerous today.

*Aldabra Atoll - a low line on the horizon*

My room at the Station looked out towards the reef, over violently green–and–blue striped water. A sunbird suspended its nest from a wire above the veranda, the busiest and noisiest place it could have chosen. Twice the nest fell down and was restored. Once, a chick tumbled out and was replaced. Then both young birds fell out, were put back and duly fledged. Reef herons, crab plovers, grey plovers, whimbrel and turnstones pottered amongst the flat heads of coral. In March a bar-tailed godwit and a sand plover, both of them resplendent in breeding plumage, turned up. Behind the tiny settlement the green Pemphis scrub was full of sunbirds, doves, white-eyes, bulbuls, drongos, pied crows, Seychelle fodies and couchals. It was never quiet for long and the sky was never empty. As a change from the usual frigatebird/booby duels there was the captivating sight of a black kite being mobbed by tropicbirds, which I never could have imagined, or a pied crow chasing a reef heron. The crows, as ever, were bold and opportunistic and more than a match for the kites. In a squabble over offal one crow tipped the kite onto its face by the judicious use of leverage, whilst its mate nipped in and stole the goods. All this provided wonderful material for my friend John Busby, who had come with me at his own expense.

*Crab plovers visit Aldabra from East Africa*

*Sunbird*

The Research Station on Aldabra fairly hummed. David Bourn might be working outside, covered in gore as he carved up a giant tortoise. He had already marked more than 5,000 on Brastakamata. The South Island is quite open, with close-cropped sward which the tortoises graze like sheep. Jenny Wilson, working on the flightless rail, an endangered species, might be coaxing her entrancing chicks (hatched in incubators to save them from likely predation) to feed. They were tiny balls of fluff delicate as thistledown. One of them refused the proprietary food and began to drink a lot of water so I suggested force-feeding with more natural food, like fragments of crab dipped in sea-water. To our delight it began to eat ravenously and was soon thriving. Nothing could be more endearing mites than these fragile rail chicks. Jenny dug up ant pupae for them but these delectable items came mixed up with live ants and detritus from the Casuarina trees. The baby rails loved the juicy pupae but were wary of the biting ants. They cautiously approached the wriggling pile and proceeded to stuff themselves only to leap high into the air, kick out vio-

*Rail*

188

*Pied crows and black kite with a dimorphic egret*

lently to dislodge an ant, fall flat on their backs and scurry off to a hiding place before venturing out again.

The ancient and prestigious Royal Society of London was doing its impressive best to run a varied research programme on Aldabra but it was fiendishly difficult, beset by awkward logistics for supplies and personnel, lousy communication (just a faint and highly erratic radio link via South Africa) and frequent labour problems in the form of strikes and quarrels. Apart from the scientists and staff the only other inhabitants were the Seychellois workers and their families. The Royal Society had built an Aladdin's cave stocked with treasures you would never dream of: Johnny Walker's Black Label (perish the thought of mere Red Label), Dundee cake and tinned salmon. The lowly sardines that had sustained me and June for a year in the Galapagos fell abysmally below Royal Society standards. Unbelievably, the library was air-conditioned and the librarian wore a bikini. June at least had the edge there: she hadn't needed a bikini, but then we hadn't got a library. There was a laboratory, workshop, dark-room and communal dining room. The rain-water tanks held 58,000 gallons and their solar panels provided hot water. It was hard to imagine an establishment of that size and complexity in mid-ocean with no access by air and organised from halfway across the globe. No wonder it was a drain on the resources and administration of even the well-endowed Royal Society. It was not to be expected that research and admin. would proceed without a hitch but by and large it worked amazingly well.

Aldabra was my first experience of working alongside others as part of a small, isolated community. Previously, however remote the location, I had always organised my fieldwork myself and carried it out with June's help, free of complications. Aldabra was very different. During my time there other people were working on projects as diverse as plant distribution, giant tortoises, flightless rails, nematodes in leaf-litter and the breeding ecology of red-railed tropicbirds. Becky Wodell was battling with Aldabra's coucals. She wanted to mark one member of a pair to see who did the nest-building and incubation but she could not catch them for fear they would desert. So she stuck a glass tube through the roof of the nest, attached some plastic tubing and hoped to blow a tiny amount of paint onto the sitting bird. The coucal had other ideas. It entered the nest, picked up the glass tube and tilted it. The paint dribbled out and when Becky blew there was a rude noise but no paint. The coucal emerged, looked around and went back in, still unmarked. Maybe it was this episode that, a year or two later, gave me the idea of using paint to mark frigatebirds. Barry Reville, my Australian Ph.D. student, built a hide in the colony and using a blowpipe fired small, paint-soaked pellets at the frigatebirds. The paint made just enough of a mark to clinch identification, which in frigatebirds is devilishly difficult but absolutely essential. Without it you are sunk, but catching frigatebirds high in mangroves is about as hard as it gets besides risking damage to the birds. And it would certainly distort results by leading to desertion or take-over. Some years earlier (in my high-tech phase) I had used a blob of green paint applied at the end of a long pole to mark Abbott's booby and this 'technique' was later modified to mark frigatebirds.

To carry out all these projects the Station's personnel had to cross several kilometres of lagoon to visit far-flung parts of the atoll. Often, they were away for days at a time, taking food and equipment and one of the Seychellois, usually Harry, as boatman, cook and handyman. If several people wanted to be away from base simultaneously the organisational demands dwarfed British Rail's problems. The run on boats and outboards was continuous, posing big problems for engine maintenance and fuel supply. To cope with this the dour Aberdonian mechanic, who never, in his wildest dreams had imagined himself so far from Union Street, needed frequent recourse to a dustbinful of home-brew, which he tended devotedly.

On top of all this the expeditions across the shallow lagoon were completely at the mercy of the tide, which further restricted the number of days available for such trips. Moreover, within the lagoon itself tidal movements were complex. Some areas lagged several hours behind others because of the complicated structure of the atoll. So it was a nice exercise working out who could go where, and when. This onerous task fell on the Director, Roger Hnatuk, a botanist, and he arbitrated skilfully between competing claims for boats and helpers but the awkward logistics greatly fragmented the research. Compared with my previous experiences on the Bass, in the Galapagos and on Christmas Island of actually living full-time with my study birds, the Aldabra set-up, though unavoidable, was pretty hopeless.

*The Aldabra coastline, Indian Ocean*

*Aldabras' rocky coastline*

*Mangroves and palms*

*Islets, some undercut, dot the Aldabra lagoon and provide nesting places for seabirds*

For a few days after our arrival the tides were low and there were no trips across the lagoon. When they became high enough Chris Huxley, who was working on the flightless rail, planned a visit to Middle Island where there was also a large colony of frigatebirds. So my first trip was to Middle Island. 'Palm Hotel' was a frond-leaved hut on a lonely coral beach, which to me felt quite like old times. The indispensable Harry, stocky and powerful, was acting nanny and it was no picnic for him. We had barely beached the dory before he disappeared with make-shift line, hook and crab-bait to return an hour later with three fine fish for supper. They went perfectly with rice and curry (Harry again). It seemed a good idea to turn in early on our primitive wooden bunks but the peaceful night, inky black, was soon rent by wild cries from John in the throes of a nightmare and then by thumps as he terrorised the foraging land crabs by walloping the ground with his shoe. Since he operated within inches of my head I would have preferred the crabs. There were also rats, towards which John's hostility proved commendably impartial, so we had quite a musical night.

We crossed the lagoon in a 6m dory. Middle Island is a maze of mangrove creeks, alive with rays and small sharks. Even a small shark can take a sizeable chunk out of a

*Fidler crabs*

leg as Tony Diamond discovered when wading in little more than two feet of water. The sharks were mainly white-tips, the very same species which had been abundant in the Galapagos where, with no danger at all, we regularly swam through dozens to reach our dinghy. Yet here in Aldabra it was risky even to wade. I don't know how such variability is to be explained.

Early next morning we penetrated deep into the colony of frigatebirds, the sky alive with their effortlessly soaring forms, forked tails scissoring. They nested in clusters which had grown out of the earlier clusters of displaying males. Between the clumps were large empty areas. There had been massive changes since Tony's survey two years previously and areas which had been densely colonised were now empty whilst new groups had appeared. These changes didn't surprise me, knowing how frigatebird nesting groups arise, nor were they difficult to map, but how best to study frigatebird behaviour was going to be a tough nut to crack. The nests were up to seven metres high in mangroves which even at mud level were devilishly difficult to negotiate and I was soon convinced that the only way to get to grips with their home life would be to live amongst them for days at a time. But how can you live in a mangrove swamp? The answer seemed to be to build a hide and live in it. Nobody had ever done that and no wonder for it would be uncomfortable, stressful and even dangerous. The Dutch ethologist Adrian Kortlandt, eccentric but brilliant and later renowned for his work on chimpanzees, came close to it in his early and classical study of cormorants, which he carried out from a 'live-in' tree hide. When Barry Reville eventually did it on Aldabra he got fascinating results. Helping to make this study possible was the best thing to come out of my visit. And later, on the strength of this study, Barry was chosen to head the research on Abbott's booby on Christmas Island, my particular interest. So luckily it all paid off.

A big snag with frigatebirds is that it is next to impossible to recognise individuals. Our efforts on this score in the Galapagos had been risible. We had ended up by sticking bits of elastoplast on their bills, hardly cutting-edge technology especially since they fell off. We would have used paint-marks but we had no paint. The legs are so tiny, and hidden in the belly-feathers, that rings are useless. Wing-tags are possible but seemed a bit suspect on such a stupendous aerial acrobat. Wouldn't the stainless-steel pin tear away from the leading edge of the wing, secured as it was through the membrane? Tony Diamond had used them successfully, each tag boldly numbered with black-on-yellow but I wasn't keen and in any case catching and handling displaying birds was incompatible with a study of their undisturbed behaviour, which I deemed essential. Marking them with paint-soaked pellets fired from a hide, using a blow-pipe, left the birds completely unstressed.

*Male lesser frigate*

As I said, boat trips to various parts of the atoll were complicated by tides or other things but I was still interested in earmarking groups best suited for further study, so on March 11th I paid another visit to Frigate Island. This time, some earnest devotee of lateral thinking had the brilliant idea that the launch should go outside the lagoon and that I (and anybody else in my party) would use the dory, first to travel inside the lagoon but then, ah then!, we would rendezvous with the launch in the open sea. In order to do this we would have to pass through a gap called Johnny Channel, where the tide races at a vicious six or seven knots. By doing all this we would be free of tidal restraints in getting the launch across the shallow lagoon. But there were big snags. Had the dinghy's outboard failed, which it often did, we would have been smashed against the wicked fangs of the champignon by the fierce current and then ditched into turbulent, shark-infested seas maybe half a kilometre from the waiting launch.

*Young greater frigates*

Somehow the whole idea lacked immediate appeal and it certainly didn't improve with afterthought. So I was mightily relieved when Chris Huxley, an experienced yachtsman, scuttled this harebrained scheme before it scuttled us. Talking of scuttling, on this trip a curiously bent waterspout twisting far up into the sky and fearsomely turbulent at its base, moved across the sea. I wouldn't have fancied being underneath that. At this season some of the deluges of rain were so heavy that you simply couldn't move. Anybody caught in one could only sit it out, like being under a water-fall.

Meanwhile that miserable old crate 'Nordvaer' had limped into Mombasa with desperate engine trouble and was awaiting spare parts from Germany. Our official schedule was shot to pieces and we had no idea when we would get off the atoll since 'Nordvaer' was the only vessel that ever called. It was more than ironic. I love tropical seabird islands and our year in the Galapagos had been incomparably more spartan and difficult than Aldabra, with its air-conditioned library and Johnny Walker's 'Black Label'. But our time in the Galapagos had been planned and there had been no competing claims on time and emotions. Aldabra was a totally different kettle of fish. My expedition on behalf of the Royal Society was meant to be brief and as things had turned out it was a tricky time for me to be away, squeezed perilously into a gap in my University teaching and with newborn twins at home. Few people even knew that I had gone! So, sadly, Aldabra was wasted on me. As time passed and our radio links with London proved tenuous and unreliable, it became all too clear that the august Royal Society had no fall-back plan, Aldabra began to feel like a rather wonderful prison. There was no way of sending or receiving mail and in the absence of guidance from the Royal Society we were well and truly marooned. As things turned out it was almost six months before Nordvaer got back to Aldabra and in all that time there had been no alternative relief boat.

John seemed admirably nonchalant. He was having a wonderful time producing wildlife drawings and paintings which he later exhibited in London and some of which adorn this book. But once I had sussed out the frigatebird situation, which I did pretty quickly, I had little more useful work to do. There were plenty of fun-observations to be made but I wasn't in the mood. I felt distinctly tetchy, which was perhaps why I was a bit argumentative. I love John, who is generous, creative and caring, but we have very different approaches to things. On one occasion he insisted that art is entirely – note 'entirely' – logical. A picture is as precise as an algebraic equation and the viewer, if he understands enough about art, can know exactly how that picture should be composed and what the artist meant to convey to the viewer. In the case of modern art, I'm not sure that the artist always knows! But what about pickled sheep, unmade beds, dung heaps and screwed up balls of paper thrown out of an aeroplane? What is art?

Providentially, we were not forced to wait for wretched 'Nordvaer' to stagger out again or, on my return, I might well have found myself bereft of job, wife and children. At 14.00 hrs. on March 18th 1974, quite out of the blue, an odd-looking schooner called 'Argo' coasted along the edge of the reef and dropped anchor. She looked a bit weird because she had been sawn in half and a bit slotted in to make her longer. No matter; dear old 'Argo' stands high in my affection, alongside 'Pacific Queen' who took us to Peru's legendary guano islands: 'Cristobal Carrier', the unseaworthy hulk in which we slipped down the muddy estuary of the River Guas one velvety midnight, bound for Academy Bay in the Galapagos: 'Lucent', the straight stemmed old wooden yacht in which skipper Dave Balfour shipped us, our ten wooden chests and a year's supply of food in motley cardboard boxes, to uninhabited Tower Island: 'Beagle II', skippered by charismatic Carl Angermeyer, which transported us across the equator three times in less than a year and dear old 'Hoi-Hoouw' which plied the Java Straits between Singapore and Christmas Island. May their names endure for few, if any, are still afloat. But Roger, who buzzed out to 'Argo' in the station longboat, returned with dire news. 'Argo' could not take passengers! She had been unable to sign on a European mate before leaving Mahe for Mombasa and with only the skipper and a young Seychellois crew she had been forbidden to carry passengers on the grounds that, should anything untoward happen to the captain, such as a heart attack or falling overboard, there had to be somebody else on board capable of navigating. This was taken so seriously that the passengers already on board at Mahe had been obliged to disembark, which can't have pleased the Plantation Manager on Assumption Island, whose wife was amongst them! A warning telegram had been sent to Aldabra to thwart us but mercifully it had not arrived. That simple slip may have altered the whole of my subsequent life. It was an unexpected blessing of the poor communications which we had so often cursed. For one breathless moment it seemed that a heaven sent chance to get to Mombasa had been cruelly scuppered and that little 'Argo' would sail away leaving us gibbering with frustration. But, bless him, the laid-back young South African skipper, a fine, fine breed, took pity on us. I eagerly signed a declaration that I was a competent navigator and, though I am not even an incompetent one I probably could have hit Africa, perhaps literally, and the job was done. Out

*The "Argo" which rescued us from Aldabra*

of my many island leave-takings this was the sweetest, a moment to savour. I can still feel it. Fifty giant tortoises were hoisted aboard – they were the reason why 'Argo' had called – to join the 14 tons of Assumption guano which had been partitioned off. We boarded and the welcome sound of anchor chain in hawser-pipe signalled our imminent release from a paradise whose only fault had been in me and the timing.

The 23m 'Argo' belonged to Harry Saville of the Seychelles who held a special licence to export giant tortoises from Aldabra for distribution to zoos. These had been collected well before 'Argo's' arrival and penned at the station and all that remained was to manhandle them into 'Fanny', the Station's long-boat, and transfer them to 'Argo's' hold together with a boat-load of fodder, though this provender was kindness rather than necessity. Galapagos tortoises have been known to live for a year without food, a capability which whalers visiting the Galapagos both knew and took advantage of, so a few days would hardly have bothered the Aldabran variety.

*Tortoises & cargo being loaded into the station long-boat to be taken out to MV 'Argo', anchored outside the fringing reef*

'Argo's' post-operative scars from the insertion of a mid-section to lengthen her, were still clearly visible. The view for'ad from the wheel, which was positioned far aft, was totally blocked by the rear wall of a large cabin which straddled the entire width of the boat, a failing which was to cause the skipper a deal of anxiety when he eventually had to negotiate the buoys and traffic of Mombasa harbour. 'Argo' had a jaunty bowsprit, a mast and boom made from Casuarina wood, a squareish mainsail of heavy cotton and a mizzen. After sailing from Mahe on September 12th she had called at Cosmoledo and on March 16th had arrived at Assumption Island. She had been scheduled to make Aldabra, some 13 kilometres away, the same day, but it happened to be a Saturday and everybody on Assumption was drunk. What with that, and the need to build a wooden partition for the tortoises, she did not get away until the 18th. If ever I return to Aldabra, which is highly unlikely, the plaintive call of the lovely crab plover will not seem so melancholy and the regular crashing of the Indian Ocean breakers will soothe rather than fret, but at the moment of my departure by far the most musical sound in the world was the throb of little Argo's 87H.P. Lister engine and the creaking of the shrouds as she ploughed sturdily across the tropical Indian Ocean at a steady seven knots.

We ate on deck, there was nowhere else, sitting on the hatch cover or on tyres. From the cookhouse, a tiny wooden sentry box with 'PARADISE HOTEL' in fancy golden script above the door, came the fragrance of burning mangrove and spices. Night had now fallen and the Southern Cross glowed in the starlit sky. The ship's 'boy', a delicately handsome Seychellois, came on deck with delicious soup in fragile china cups which he placed reverently on top of a rusty old diesel drum covered with a fair linen cloth. Then came chicken, potatoes, peas and vegetables fried in butter and garlic. It was better fare than we'd had on Cunard's 'Queen Elizabeth'. The tall masts swayed, the sails bellied gently, the engine throbbed steadily and we slid contentedly westward through an oily swell; how rarely it happens – sheer perfection. A light on the horizon marked the passage of a big ship for we were now crossing the tanker route from the far east. Many a small craft has been smashed to splinters by one of these steel monsters, criminally without a lookout as it ploughed heedlessly onward, and never even felt the impact. Our skipper had been a mate on such a tanker and knew from experience that before entering Hong Kong harbour they routinely inspected their bow anchor for the tell-tale remnants of Chinese junk-sails so that they could remove the incriminating evidence. It seems unbelievably negligent but almost without exception voyagers in small boats tell of terrifying near-misses from large ships rushing heedlessly through the night. Just as in the filthy habit of jettisoning oil at sea, it is impossible to nail the criminal.

Next day a rusty old tanker, the 'Schneldt' from Rostok, Russia, altered course to inspect our ragged little schooner with its four sails, two wooden hatches on deck, the shrouds festooned with drying fish and its crew of ten Seychellois in wide-brimmed straw hats. And they couldn't even see the fifty giant tortoises! Meanwhile the old engine thumped away as we devoured the distance to Mombasa. 'Argo' may have been a home-made schooner but she was a civilised little ship with a proper toilet, a cute tin wash-bowl with a floral pattern and a mirror. Water came from a tank with a brass tap and all was sweet-smelling. What more could you ask?

We were not the only deck-cargo. Several scrawny fowl destined for 'Paradise Hotel' squatted beneath an upturned dinghy and a poor old pig lived, firmly wedged, between the dinghy and the gunwhale. I think it was destined for Mombasa; anyway it didn't figure on our menu. Little ships like 'Argo', lacking refrigeration, have to keep food alive for as long as possible, which is not much fun for the wretched creatures.

The sea is a pasture with plankton instead of grass, but few pastures are as patchy. Some parts of the sea are richer than others in minerals, plankton, fish and squid. Seabirds forage widely looking for these areas, which tend to move around pushed by winds and currents. As we crept towards Mombasa we kept running into red-footed boobies from Aldabra, chasing flying-fish as these delightful little creatures broke surface and skittered along like flat pebbles bounced across a pond. Sometimes, seen against the sun, you would think they were birds —"shooting out of the waves like arrows, with outstretched wings they sailed on the wind in graceful curves, then falling again till they touched the crest of the waves to wet their delicate wings and renew their flight". So wrote Captain Joshua Slocum in his classic 'Sailing alone around the world' in the 1800s, the first-ever solo circumnavigation, which he made in 'Spray', built with his own two hands at a total cost of a few score pounds. Flying fish were active at night, too, doubtless escaping from underwater predators. One flew into our Skipper's bunk last night.

The male red-foot is tiny, the smallest of all the booby family and perfectly adapted for the aerial pursuit of flying fish. Boobies dive for them too. I watched one catch half-a-dozen in quick succession. Several times we came across parties of hundreds of sooty terns, endlessly on the wing. These attractive tropical terns rarely settle on the water

*Red-footed booby*

and it is usually stated that they live for days, weeks or even months perpetually aloft. I find this more than difficult to believe. If they are waterproof it just doesn't make sense. The ones we saw probably came from the large colony on Bird Island in the Seychelles. The biggest sooty tern colonies or 'fairs' are said to contain millions of birds but few if any are so big today. Most tern species fish close to the nesting colony and carry fish back one at a time in the bill, but sooties are different. They forage perhaps 100 km or more from the colony and it would be uneconomical to carry one fish all that way, so, like boobies, they transport their catch in the crop and stomach. Appallingly, many sooty tern colonies are devastated by cats, perhaps the most pestilential mammal ever introduced to seabird islands, though it faces stiff competition for that accolade from rats, whilst in a few cases pigs, goats, dogs and even mice are no slouches either. The damage to seabird colonies has been truly incalculable, another of man's heavy footprints. It may be no wild exaggeration to say that, one way and another, man has reduced the world's seabirds by 90% or more.

*Red-footed booby greeting display*

*Great frigate sunning itself*

It was endlessly pleasurable in the warm sun on a benevolent sea, watching seabirds, dolphins and whales but the nights were a bit nerve-racking. It takes time to get used to surging heedlessly through the inky darkness. What if a heavy baulk of timber or some other massive debris lurks ahead, loitering with intent. I imagine that solo sailors, at least when crossing shipping lanes, must simply resign themselves to fate; they have to sleep some time. John was far more concerned about cockroaches after one crawled through his hair. He didn't even draw it.

By day three after leaving Aldabra we were nearing the coast of Africa. A masked booby in its first year, still brown rather than the dazzling white of the adult, came to inspect us. It could have wandered thousands of kilometres from its birthplace in the Red Sea, Gulf of Arabia or perhaps Latham Island off the African coast. Thousands of adult seabirds that, for one reason or another are not engaged in breeding, wander the sea-lanes. For instance, Clipperton Island in the Eastern Pacific was home to a huge colony of masked boobies until pigs got ashore, multiplied and ate their eggs and chicks. Most of the adult boobies then deserted the island. An admirable American called Stager shot the pigs and within a year or two there were once again thousands of boobies back nesting on the island. Where did they all come from? Were thousands of potential breeders roaming the seas around Clipperton with nowhere else to nest, and Clipperton pig-infested? It seems unlikely, and yet they re-appeared as if by magic once the pigs had gone. What would they have done had Clipperton remained infested?

Africa showed up at 2.30 p.m. on March 21st and it was Africa burning. We could smell it far out at sea. As dusk drew on the lights of Mombasa harbour began to blink. At that moment, in the busy approaches to a major seaport, the drawbacks of 'Argo's' idiosyncratic design became alarmingly clear. Andrew, our noncholant young Skipper, urgently needed a clear view of the intricate buoyage and the myriad harbour lights which he had to decipher but all he could see from a standing position at the wheel was the high blank wall of the cabin in front of him! So our resourceful captain sat on the cabin roof facing aft, the way we had come, whilst looking over his shoulder at the way we wanted to go, and twiddling the wheel with his bare toes. Slocum would have been proud of him. A powerful harbour launch surged out, fixed us in the mesmerising beam of a searchlight and yelled "Who are you"? "We are MV 'Argo' from Seychelles". "O.K we are coming aboard". And come they did, bringing their confounded launch with them, for their fenders were too high for poor little 'Argo' and their gunwhales – of course well protected – stove in our wooden dinghy which was stowed on deck. Did they apologise? Did they even mention it? No and no. That is so typical of arrogant officialdom the world over. They should have been charged for the damage, as they would have charged a guilty party, but nothing would have come of it. No wonder there are anarchists here and there.

As we entered Killindini Harbour what did we see but wretched 'Nordvaer' lying smugly alongside a couple of Japanese fishing boats as though she had all the time in the world. This was March and she didn't move until June. The forest of boats, derricks and buildings must have stunned those simple Seychellois who had never before left the island of Mahe. Morning light revealed black-headed and Aden gulls, common terns, kites and white-rumped swifts but the Africans were far more interested in the dried fish festooning 'Argo's' rigging. They bartered eagerly with the crew and doubtless got a fine bargain from the simple Seychellois.

Little more than 24 hours later on a bitingly raw day I touched down in Aberdeen that I had left on a bitterly cold and snowy January day. It felt unreal, so soon after Aldabra. Is it the abrupt annihilation of huge distance, the harsh cultural shifts or the fatigue of air-travel that unfailingly brings me home with a dismal, heart-sinking sense of anti-climax, almost of failure? It is odd and seems not to apply to sea-travel. But on top of that, and alone of all my trips, Aldabra lingers uneasily in my mind and will remain so. Unfinished business and lots of question marks.

*Red-tailed tropic birds*

*Cape Kidnappers*

## Chapter 11

# TE-MATU-A-MANI; CAPE KIDNAPPERS

*"Observations in biology have probably produced more insights than all experiments combined"*

Ernst Mayr 2002 'What Evolution is' Weidenfeld & Nicolson  London

The North Sea or 'gannet bath' of a sixth century Anglo-Saxon poem is grey and turbulent, swirling against the unyielding basalt of the Bass Rock, the classical breeding place of the Atlantic gannet. They haunt its ledges, soaring along the dark cliffs and circling in mounting columns on the updraughts that bounce from the sea in the lee of the Rock. Their strident voices, like rusty old anchor chains dragging through iron pipes rise above the endless surge of the sea. The old Bass is usually cold and windy; a balmy day is an unexpected bonus but the Atlantic gannet easily copes with foul conditions, lumpy grey seas smoking with spray, icy gales and snow-encrusted ledges. The climate on an exposed Rock can be truly vile. Yet the gannets return in January to sit on hostile granite ledges, forfeiting the limited winter daylight when they could have been foraging. Nor do they merely lose valuable fishing time; they use precious energy flying hundreds of kilometres to and from their fishing grounds when they could easily have stayed safely at sea, their real home. Why do they do it? The answer is simple: to reclaim their nest site before somebody else takes it over.

On the other side of the world, at the southern tip of Hawke's Bay in New Zealand, lies Cape Kidnappers, the most famous home of the Australasian gannet, much like our's but with more black on wings and tail. Cape Kidnappers looks a bit like the Seven Sisters of the white cliffs of Dover. Made of smooth, silver-grey mudstone, it gleams in the clear air and New Zealand sunshine. Erosion has scooped out the North face until it is now concave and dangerously overhung. The final outpost of the Cape which long ago lost its umbilical connection to North Island rises from the sea like a monster bee-hive. The top of the mainland cliff, deeply scalloped, holds in the final, most seaward crenellation more than a thousand pairs of gannets nesting with slide-rule precision, beak-thrust distance apart. Seen from the North Cape, Kidnappers curves like a giant fish-hook; the Maori name for it 'Te-matu-a-mani', man's

*The beach
approach to
Cape Kidnappers*

*Other views of
Cape Kidnappers*

fish hook with which he pulled the North Island out of the sea. It is the next best thing to an island and the gannets love it. North of the Cape the ancient cliffs are superbly stratified and slashed by tremendous ravines. In 1769 a certain Yorkshire draper, Thomas Cooke, the pride of sea-faring Whitby, charting the north-east coast of New Zealand, narrowly thwarted an attempt by the Maoris to kidnap one of the ship's boys, hence 'Cape Kidnappers'. Hawke's Bay itself commemorates Sir Edward Hawke, First Lord of the Admiralty.

No matter how you approach it Cape Kidnappers is spectacular and it must surely, also, be the most accessible gannetry in the world. You can drive right up to it, point-blank range, and although, compared to an island, it seems horribly vulnerable it remains amazingly unspoilt. The overland approach by 'Safari' which started in 1978, the year we arrived, crosses a huge sheep station, climbing precipitous valley sides with hair-pin bends, but the shortest way is along the beach from Clifton by trac-tor-trailer at low-tide. This route passes Black Reef Rocks, about eleven flat-topped Rocks crowded with gannets, but it doesn't take you to the main Plateau colony. For this you must clamber several hundred feet up a rough, steep track from sea-level. Thousands of visitors puff and pant their way up, nuns in heavy habit, school children and even bikini-clad girls on horse-back. I never saw anything like it on the old Bass; Scottish weather saw to that. Beyond the Plateau colony there is another large congre-gation, forbidden to visitors. We approached it along a dangerous knife-edge ridge with a sheer drop on each side, a bit of a high-wire act without a safety net.

*The winding path up to the Plateau Colony, as it was in 1978*

In 1978 Kidnappers was looked after by New Zealand National Parks and Reserves in the person of an émigré Scot from Dumfries. Ron Fisher was an admirably stout bulwark against the threat of commercialisation, which would fain have laid its grubby hands on the Cape and no doubt still would, if it hasn't already. Ron had no truck with tarmac paths, hand-rails, 'Health and Safety' notices and ice-cream merchants. To hell with the lot of them. They would 'empty the rainbow'. The steep zigzag path up from the beach was cut out of the cliff-face and the visitors, breathless but victorious, finally hoisted themselves onto the Plateau and came face-to-face with the dense and clamorous colony. For those unable or unwilling to puff-and-pant there was always the long overland route by land rover, or the solace of the Black Reef colony on the shore. Down on the beach a simple rest-house offered shade for a picnic, serenaded by the sweet song of the modest little grey warbler whilst skeins of snowy gannets soared overhead. That's the way to do it. Too many facilities and 'Interpretation' boards ruin the atmosphere. People are not as soft and daft as many local authorities.

*A typical group of viewers at the Plateau Colony, Cape Kidnappers*

*Part of the Plateau Colony, Cape Kidnappers 1978*

*Australasian gannet mating*

*Australasian gannets with a chick*

The best days of my life have been on small islands – the Bass, the Galapagos, Christmas Island and a dozen remote Scottish islands. Their magic, especially if uninhabited, lies in the thrill of a small, enchanted world, your own little world, a kind of escapism. Cape Kidnappers was almost an island, with the same aura of isolation. Instead of a boat for getting around we had a canary-yellow motorbike, courtesy once again of Ron Fisher and 'Lands and Survey'. What would we have done without them both? It was a modest little 125cc Yamaha with a buzz like a demented hornet and it carried the four of us, all at once, along the beach to Clifton. Becky faced the salt-spray from her perch on the petrol tank, singing loudly and a trifle defiantly. Simon, her twin, squashed in at my back whilst every so often, with a warning wail, June slipped off the end of the luggage rack. The shopping slotted in between us. It was full throttle through the rock-pools, skirting the tide line against a background of towering cliffs and a foreground of glistening boulders and swelling sea. The bike was our only link with the outside world, apart from Neil Burdon's orange tractors. Neil, from Clifton, had a fleet of museum pieces, rife with rust and corrosion, with which, at low tide, he towed a string of flat-bottomed trailers full of visitors to the Cape. Now and again the wheels sank deeply into the soft sand and everybody piled off whilst Neil extricated himself. When he turned the snub orange nose into the surf to detour a rock the tractor became amphibious, shouldering the spume aside. The visitors loved it.

*Regular trips from Clifton took visitors past Black Reef rocks and the nesting gannets*

The beach route passed a small colony of delightful white-fronted terns nesting on a large rock near Black Reef. In October they were courting, the males flying overhead with silver fish drooping from their bills, an offering to the female. It both feeds her, helping to fuel her clutch of eggs, and may perhaps allow her to assess his quality as a potential father. It could be that early humans had a similar system way back in our biological past but in modern society the substitution of money for hunting prowess, or if not money then other indicators of a man's material worth – say cattle or camels- has wrecked the old system which had directly linked physical attributes to fecundity. Men can make piles of money by cheating, lying and getting others to do their dirty work, and in a thousand underhand ways which could not have worked in a primitive society. Alas for the terns, not a single chick was reared. People stole the eggs, disturbed the birds and paved the way for predatory gulls; a familiar story.

Our route to Cape Kidnappers had been long and fortuitous. Many years before we set eyes on it, a teenage New Zealander had been smitten by the gannets, which he visited most weekends, sleeping rough. Chris Robertson, who was to do more field-work on New Zealand gannets than anybody else, wrote to me about his gannets. I have no idea what I wrote in reply but it must have been mildly helpful for, long years afterwards he suggested that I might like to come to Kidnappers myself. By then he was the Ringing Officer for the New Zealand Wildlife Service and by lucky chance I was due to take a six-month sabbatical from Aberdeen University, to do some research. What about a comparison of the behaviour of the Australasian gannet with ours? Even better, Chris wrote —"why don't you act as Ranger for the Cape gannetry whilst you are there?" It was a brilliant idea so far as I was concerned and I jumped at it. The

*New Zealand dotterel*

Ranger had the use of a cottage at the Cape which would make daily observation easy – just like living on the Bass. Only later was it diplomatically pointed out to me with typical Kiwi kindness, that the Ranger's job and the cottage was not actually anything to do with him. But that was Chris all over; a born organiser, generous and full of entrepreneurial confidence. His life is a tale of 'can do' with the occasional 'I think' tagged on. Once we had got ourselves to Napier he quickly organised supplies, piled them and the four of us plus his tiny chi-hua-hua into his capacious Holden estate and drove us out to Kidnappers by the precipitous overland route. At times the car's bonnet reared so far skywards that the track literally disappeared beneath us, like taking off in an aeroplane. Luckily it was dry. In wet weather the track becomes literally impossible because the wet top layer coheres and slides over the dry dust beneath like a layer of matting on rollers. That might be quite fun on flat ground but in some places the track slopes sideways towards an unfenced edge which drops sheer into a tremendous ravine. On the hills surrounding the Ranger's cottage John Neilson ran thousands of sheep in glorious seclusion. All of them had dirty back-ends, a consequence of the rich spring grazing.

I had six precious months in which to get to grips with the behaviour of the Australasian gannet. They were so tame that all I had to do was sit at the edge of the colony armed with stop-watch, camera, cine recorder, notebook, binoculars and – vitally important- the right questions. It had to be almost entirely a study of behaviour because ecological work would have entailed intruding into the colony to weigh eggs and chicks and, quite properly, this was out of the question. It was absolutely essential that nobody should destroy the trust and tameness of the gannets otherwise the constant stream of visitors would have wreaked havoc. As it was, they were confidently and amazingly blase.

With my usual luck, I fell squarely onto my feet at Cape Kidnappers. The Ranger's cottage, made of pine planks 5cm thick, perched on the edge of a cliff looking down the coast to Black Reef Rocks whilst the main Plateau colony was just up the hill behind us. Almost literally we could step out of the door and into the gannetry. My return for this unprecedented hospitality and research opportunity was merely to act as honorary Warden, looking after the colony and answering questions from visitors who must have wondered what this interloper from the Northern Hemisphere was doing there. I admit that I did wonder if the hordes of people would spoil everything but the tides rendered their moon-linked services once again by cutting off the shore route for most of the time. They guaranteed hours of peace during which I could watch gannets undisturbed, for the overland safaris were few and soon gone.

*Rangers' cottage, Cape Kidnappers*

Gannets down under, though much like our own, are by no means identical in behaviour and nor did I expect them to be for they have been distinct for more than a million years, but I found one intriguing difference in their fledging behaviour. The young Australasian gannet takes a much more relaxed approach to the tricky business of launching itself into the wide world. It is slower to develop and is far more casual about cutting the apron strings. The fledglings eventually wander amiably to the fringes of the colony, perhaps joining up with a small group of other youngsters, but they are in no hurry to throw themselves over the cliff in the desperate do-or-die manner of their Atlantic cousin. They exercise their wings, joust tentatively with other juveniles, preen a little and may try a short but safe flight on land before going back into the colony where they may be fed before trying again. Their Atlantic cousin's savage baptism may entail a hazardous passage through massed ranks of hostile adults before it hurls itself over the precipice and battles on untried wings against the unruly air-currents that can so easily and fatally sabotage it. There can be no return, no second chance. From the moment it jumps it is on its own, facing a tremendous challenge. Not only must it move quickly out of British waters but it must also learn to catch fish – not the easiest of prey. It doesn't take much imagination to realise how difficult that must be for a plunge-diver like the gannet. It can't be easy at any time but in rough

*Young fully grown but still un-fledged gannets at the fringe of the colony at Cape Kidnappers*

seas it must be well-nigh impossible. It should surprise nobody that very many of these unschooled youngsters starve in their first few weeks of independence. Things are much easier for the tropical boobies which are fed for many weeks or even months after they can fly. The thing that struck me most forcibly about the Australasian gannet's fledging procedure was that it fell half-way between the Atlantic gannet's harsh baptism and the less extreme procedure of the boobies.

When the Kidnapper's youngsters finally do fledge they face a hazardous journey across the Tasman Sea to Australian waters. Like their Atlantic cousins they cross more than 3,000 kilometres of ocean before reaching their destination. A few, whose migratory urge has still not been fully assuaged by the Tasman crossing continue westwards beyond Australia, occasionally turning up at St. Paul's Rocks in the Indian Ocean or even reaching South Africa where one or two have been known to breed by mating with their African cousins. In much the same way the young African gannet migrates northward and if it overshoots its normal range may even meet up with Atlantic gannets north of the Gulf of Guinea. So far, despite claims of African gannets in waters close to Britain it is not known ever to have bred with our gannet. It would be a much bigger challenge for an African or Australasian gannet to mate with an Atlantic than for them to interbreed with each other. The Atlantic gannet really does differ from the other two in several important ways.

It may seem odd that gannets bother to migrate at all. Usually birds move out of their breeding area because they couldn't survive winter there. What could a cuckoo or a nightjar find to eat in wintry Scotland? Even a lot of hardy Scottish birds perish! But whereas winter may rarely bother an adult Atlantic gannet, which is adept at fishing in rough seas and can survive for weeks on its fat reserves, the juvenile gannet has

*The variable oystercatcher and white-fronted tern*

a lot to learn and still has to perfect its highly skilled hunting method, to say nothing of acquiring the ability to interpret the signs which may indicate fish shoals, knowledge which comes only with time. Wide experience of weather conditions and their relationship to fishing opportunities may also be important.  So in the young gannet's case, and this presumably applies to the New Zealand birds as well, a warmer climate and calmer seas may well mean the difference between life and death. Even so the first months of independence are a severe test which many of them fail. Perhaps only 10 % of the year's crop of youngsters will survive to breed. By then the others of its year will be long dead, consumed by the sea or rotting amongst the tide-wrack on some lonely coast. It is the Royal Road to perfection, the weeding out of the imperfect by natural selection and of the merely unlucky by ill-chance. "So careful of the type she seems, so careless of the single life".

*White fronted tern with fish*

*Little pied and black shags, shoveller and distant black swans*

Chapter 12

# LIFE AT KIDNAPPERS

*"Gannets are God's own birds ….. they remain Lords of the waves, bright beings of distance and air"*

Christopher Rush 2007 'Hellfire and Herring: a Childhood Remembered' Profile Books Ltd

The sun streamed into the pleasant living room of our cottage at Cape Kidnappers – not ours, exactly, but New Zealand's Wildlife Department. It was October and the Antipodean spring weather was still unsettled. Yesterday gave us a taste of Scotland with grey skies, heavy rain and a chilly, blustery south westerly, just like April north westerlies on the Bass Rock. The green seas were white topped though my look out above the gannets at Black Reef Rocks was sheltered enough; in fact quite cosy. Becky, our four year-old was stuffed with cold, but still chirpy. Yesterday she announced "My husband works on turtles in Paradise. There's not many jobs there, you know".

I was still settling into a routine, working out how to tackle the Australasian gannet's behaviour to allow detailed comparison with out own gannet. As so often, the work was inevitably dull and repetitive. I used a video-recorder and the battery leaked and corroded its compartment. It meant more hassle. Today, Ron Fisher, of New Zealand Wildlife, helpful as ever, came out with milk, bread and a paper. We feel a bit like the Keepers on the Bass Rock when Fred Marr sailed out with supplies from North Berwick. Ron cut the grass (sit-on mower of course) and fixed a non-return valve on the water system which is now a maze of pipes and a complete mystery to me. I'd never make a plumber, though maybe I could be a house painter "if you can pee you can paint". I have a New Zealand university lawyer friend who decided he'd rather paint houses than put up with intra-departmental feuding, and is doing very well.

Neil Burden chugged along the beach with his ancient tractor and four-wheeled trailers, two or three in tandem, packed with visitors. Neil is a typical Kiwi, tough, resourceful, bluff and nasal. He gave us a lift to Clifton from where Ron took us to Napier. It seems that, willy-nilly, I've bought a pale blue, ancient Holden Kingsway. The previous owner was a solicitor in Napier. It carries a huge load of luggage which we shall need when we go on holiday to South Island.

Lands and Survey, benevolent as ever, have offered us an unfurnished farm cottage near Napier for use when we are not at the Cape. Everybody is so kind. We can easily improvise for furniture and an abandoned cable drum will serve as a table.

The ride back from Clifton to Kidnappers along the tide line on our little Yamaha was wet, in heavy rain with a near gale blowing onshore, bringing sand and spray with it. I saw my first variable oystercatcher, all black but otherwise much like our own, with the same piping voice. It waded deeply, swimming at times just as our own oystercatcher will do. Two years ago a wind like this blew more than 200 fully feathered young gannets off the Black Reef Rocks and onto the beach. A team of volunteers put them back but without any idea where they had come from. Not many would survive, but what can you do? Leaving animals to die without even trying to help is a miserable business.

*Ron Fisher*

*Black Reef rocks, each with their group of nesting gannets, the range of group size makes an interesting study (timing and synchronicity of breeding)*

The social climate in New Zealand seems to be changing rapidly. Ron says that crime has rocketed since the sixties. What was it that seems to have made the sixties so pivotal? They were certainly special for me – a period of calm before the transition to the bedlam that is modern life. The old New Zealand is firmly on its way out. It really hurts many Kiwis who come 'home' to Britain full of nostalgic excitement only, as one said, to find us offhand or downright rude and apparently more welcoming to immigrants than to them, whose kith and kin are still here.

October 23rd
Labour Day, brought showers and a gale of biting wind straight from Antarctica. Our first batch of mail came out from Wellington. There were more notebooks, a spring-balance, a welcome note from the Bank of New Zealand to say that our plan for transferring money was working (remember this was more than 30 years ago) and part of her thesis on the gannets of Ailsa Craig from my student Sarah Wanless. Labour Day always brought hordes of visitors and as a result of the disturbance some non-breeding gannets on the fringes have temporarily deserted the colony. They are always warier than the birds with eggs or chicks. Ron told me that the Board which administers Cape Kidnappers are keen to have my 'views' on the present situation. What can I say? I agree with their low key approach and with keeping the Saddle Colony closed to visitors. Maybe it would be better to move the low rope fence a bit further back from the edge of the colony and put a more detailed explanation on the 'keep off' notices – just a brief account of what happens when a gannetry is disturbed. An all-weather double-sided display board under a simple roof showing and explaining what the various gannet displays mean, would be a good idea. Later I designed one which I think was eventually used. It is not always easy to strike the balance between too much and too little information.

October 24th.
The gas fridge went out and trying to re-light it produced a flash and a sheet of flame. At the second attempt ominous wreaths of smoke hinted that all was not well; the insulation burned merrily. I dashed out to turn off the gas cylinder whilst June wielded the fire extinguisher, dowsing the flames but creating drifts of white powder. More work for poor old Ron. He must be heartily sick of us, although he never shows it. Luckily nothing similar happened to our paraffin fridge in the Galapagos or all my films would have been ruined by the heat and humidity. So much depends on luck and I seem to have enjoyed more than my fair share.

Last night a mouse fell into a half full bottle of cream. It was in a parlous state and although we washed, dried and warmed it, it was still one miserable mouse. It doesn't make much sense to set traps for mice, which we do at home, then when one tries its best to kill itself, try to save it. It reminded me of the mocking bird that fell into a jar of vinegar in the Galapagos. June dried that down her bra but declined to repeat the trick with the mouse. I call that speciesism.

There are dozens of broken eggs and deserted nests amongst the gannets of the plateau. I'm not sure what has happened but suspect gross disturbance, if not from humans then perhaps goats or sheep. I have seen sheep wander casually through the middle of the colony sending gannets fleeing in all directions. Ron says there are more sheep than usual so they may well be the culprits. One good reason why so many seabirds nest on islands and especially on cliffs is that they are safer.

October 30th

I had to go to Clifton to arrange a trip to Canberra where I am due to give evidence to the British Phosphate Commissioners about the effects of mining for phosphate on Christmas Island's population of Abbott's booby. The Yamaha picked up a puncture so I took it to Ian Hope's garage; humming with activity and crackling with radio talk, like a trawler at sea. In isolated communities, garages like Ian's are today's equivalent of the old hostelries, stables and blacksmiths rolled into one. I do like the casual, friendly, no fuss or panic of the Kiwi's attitude. "She'll be right" and 'she' usually is. Whilst Ian repaired the tyre, I wandered down to the river where black-winged stilts were displaying, parading and pivoting with slowly-beating wings and lowered head.

*Black-winged stilt display*

Nov 3rd

A cryptic letter arrived from my friend John Wells of Wellington University, who used to teach in Aberdeen's Zoology Department. Would I please contact Christmas Island! Just like that! How? Who? What about? It isn't that simple from Cape Kidnappers – at least it wasn't in 1978. I suspect that the message came from Dave Powell who is desperately worried about the mayhem that is taking place there but I can't do anything from here.

Nov 4th

A tractor load of visitors arrived so I donned my 'Reserve Ranger' shirt and went on duty. Our children were playing near the cottage. Becky; "we're making stew from sheep's piddle; do you want to come back for some when you've dumped that lot?" Visiting nun, seeing hundreds of gannets sitting on eggs, "Are those birds nesting?" "Yes ma'am". A much more interesting question would have been "why are they nesting so close together?" but I suppose that might not strike most folk as unusual. Nor would they stay long enough to take in the answer. Life, even gannet life, is complicated and most folk want simple answers. It's a trend. Less effort equals better.

Nov 11th

Yesterday I travelled First-class on Air New Zealand to Canberra. I could easily get used to this. Today my New Zealand internal clock woke me at 4 30 Australian time so it was no hardship arriving early at ANPW's office. Tony Kershaw and John Hoar, two of BPC's senior men, arrived looking sombre, clearly aware of the virtually irreconcilable conflict between mining Christmas Island for phosphate whilst conserving the jungle. After detailing plans for the proposed National Park – their 'peace offering' as they put it – we moved on to the proposals for saving some of the areas crucial for Abbott's booby from the bulldozers. I'm afraid I rather hogged the floor on this one but the whole point of my being there was to convince them that those areas had to be saved. Destruction of prime habitat would mean that fewer boobies would be produced which in turn would mean a smaller, less viable population in the future. It is just not possible that, once the population has been greatly reduced, it will somehow increase its output to make good the losses. That can happen only when a reduced population gains some advantage such as more food per individual, which then improves productivity. In Abbott's case that, quite, obviously did not apply, for its population was already tiny and its foraging area vast. Knocking a few hundred birds off the population wouldn't remotely benefit the remainder. It would have the opposite effect because it would reduce social contact and hence pairing. And when a natural catastrophe such as a cyclone wipes out 80 or 90% of the year's youngsters, as can happen, the impact is all the greater if the population is small to begin with. To be fair, Tony Kershaw did agree that they couldn't mine areas crucial for Abbott's and at the same time pursue a sound conservation policy. The phosphate reserves in the disputed areas were estimated to be around 375,000 tons which wasn't a vast amount in terms of the projected yield from the island, but he didn't have the authority to concede it and so agreed to put the matter to the Commissioners at the crunch meeting next month in Auckland.

Next day Dave Powell, Conservation Officer for BPC telephoned to Canberra with a stark message which I wish I'd had in time for yesterday's meeting. He was deeply concerned and had halted all clearing operations. Abbott's has taken a severe beating, losing habitat and suffering many casualties of breeding adults and chicks. He feared another push to clear more jungle could prove too much for this small, endangered population to bear.

The return flight to Wellington passed very pleasantly in the company of Ralph Johnson, Dean of the Medical Faculty in Otago and Wellington and co-author of a book on 'Disorders of the autonomic nervous system' Mine felt pretty disordered after a series of lousy nights and early mornings. I'm not built for stress; I'd never make a politician. Nice peaceful sessions in a hide watching gannets or a quiet life on a small island or in the desert are more my cup of tea, though I did like Christmas Island.

## Nov 14th

Three Japanese journalists from Yomiuri Shimburu, a newspaper with a circulation of nine million, visited the Cape and asked me to tell them a famous gannet fable. Nothing daunted, I told them the hoary one about executing prisoners by floating them in the sea with a herring fixed to their forehead – prime targets for the gannet's lethal bill. Complete rubbish of course. Which reminds me of another Becky-ism "Daddy, where will you die and how will we get you to the graveyard?" Not death by gannet impalement anyway.

## Nov 17th

Today we met Clive Lewis and Graham Wragge, two of Ron's part-time rangers at the Cape who would normally have wardened the gannetry but for this wretched interloper from Scotland. I could well imagine how they felt. Later both of them became our firm friends and Graham eventually completed a D Phil at Oxford on fossil birds, in between remunerative salmon fishing in Alaska and exploring the SW Pacific in his yacht having, in the meantime somehow acquired a Master's ticket. Quite a boy is Graham. More recently he became involved in setting up a small airline in Fiji. In between he fitted in a couple of land rover trips across the Sahara to dispel the tedium. One plan that did fall through amidst all this was to establish a fish farm in Russia. I don't think Richard Branson could hold a candle to Graham. And he seems just as entrepreneurial with the fair sex. The secret must be physical and mental energy.

## Nov 19th

Today Neil Burdon's tractor trips were augmented by three private tractor-trailer parties. In addition three visitors came on horseback, some on motor bikes or in jeeps and others walked. Speedboats buzzed offshore and pleasure-plane trips flew overhead. It sounds chaotic but it was less intrusive than one might suppose. It didn't start until 10am when the tide had dropped far enough to allow passage and cars and bikes cannot get up to the colony. This saves the gannets from noise and mechanical disturbance and the people who do climb to the Plateau are usually well-behaved and soon leave. So it is by no means as bad as it sounds. And on the credit side one hopes that an appreciation of wildlife will be encouraged and bear fruit

## Nov 21st

Ron showed me where pauas live, adhering to rocks around or below low tide mark. This black mollusc has a heavily encrusted shell beautifully lined with mother-of-pearl. Whilst we poked around a large octopus slithered away – the first I had seen in the wild. To cook pauas you take them out of the shell, beat them savagely against a rock and then fry them in butter and garlic for a few seconds. Said to be a dish fit for a king, they seem more suited for a long life as the sole of a boot. Some days, at least a dozen sackfuls are gathered, each containing more than the statutory 10 per sack.

A large party of delightful sixth-form girls from Hastings came out from Clifton. Some seemed keenly interested in the gannets and asked a few questions but most spent half a minute gazing blankly before buzzing off back down to the beach

Dec 12th

To Auckland to meet the BP commissioners who hold the fate of Abbott's booby in their hands, although I suppose in the last resort the Australian Government is the boss. I really don't know why they invited me. I expected a fairly formal cross-examination about the need for conservation and the whole Abbott's booby saga, but all we had was an informal chat over sandwiches in the boardroom. Commissioner Bremner, by reputation a tough businessman, said straightaway that they had agreed to a two-year moratorium on further jungle clearances, at least in sensitive areas. "So you've already won your battle without having to fight it". I had not expected the Commissioners to know much about Abbott's and the various goings-on at Christmas Island and it was soon clear that most of them didn't. They didn't even know that Christmas Island was Abbott's sole nesting place in the world. Before the meeting broke up and the commissioners got down to their real business, I asked the only question that was addressed formally to the whole board. Did they view the two-year moratorium on jungle-clearing as a green light to go ahead afterwards or was it the basis for a long-term management plan? Bremner replied, distinctly testily, that they could not anticipate what they might do in two year's time. But the whole exercise is fruitless unless it does form the basis for assessing Abbott's long-term needs. A mere two year delay before simply carrying on with jungle clearing is useless. But they have approved the plan for the Christmas Island National Park. Hooray! What a triumph.

Dec 15th

Trevor Poyser (T & AD Poyser, publishers) sent a copy of my recent monograph on the Atlantic gannet. They have done a splendid job on the production – he is a perfectionist and quite unflappable, as well he might be. During the 1939-45 war he flew gliders at night behind enemy lines in France, landing in near blackness on a pocket handkerchief. I must remember to ask him how he got home again!

Dec 16th

To 'Marineland' in Napier, where the dolphins and 'Flash' the 14 year-old sea-lion who is supposed to be a bull but looks every inch a cow to me and is clever enough to be one, fascinated the children. The dolphins jumped 5m out of the water to reach a hanging ball and nosed balls vertically clean out of the water to hit skittles on the end of the pool. Catching fish for a living evidently endows animals with exceptional co-ordination and little wonder. When all the other visitors had left the trainer allowed the children to play ball with the dolphins at the edge of the pool. Even this was trumped by Flash who juggled five sticks on the top of each other with a ball atop the lot. As if this was not unbelievable enough he then jerked his head so that the sticks fell, whereupon he caught the ball on his nose. Then he balanced a cup of water on a long stick and rolled completely over without spilling a drop. Even wild sea-lions in the Galapagos juggled with empty shells when there was nobody to applaud them except the odd lava gull or perhaps an ancient iguana basking in the sun. Like cetaceans and other carnivores even adult sea-lions seem to enjoy play. A high protein diet means that, unlike herbivores, they have lots of spare time for play, which may help them by sharpening their prey catching and handling skills. Young gannets juggle sticks and feathers using the exact movements needed to handle and position a fish before swallowing it. That is not co-incidence, it is adaptive evolution.

## Dec 18th

A red letter day. Soon after we got back to Napier Chris Robertson arrived. "Right" he said "you're flying tomorrow". And indeed I was, starting at 6 45 from Napier Airport. The pilot ("Be with you in a mo") a spotty-faced youngster in slacks, shirt and tie, roared up in a Morris Minor. You see a lot of lovely old British cars in New Zealand. The climate suits them, just as it does me; they don't rust.

We took off in a four-seater Piper Cherokee on a glorious morning, out over Hawke's Bay and north to Portland Island. The ride was bumpy and the steep banking manoeuvres, necessary to get the wing out of the way for Chris' photographs, did nothing for early morning stomachs. The horizon went crazy, but to my relief I experienced nothing worse than a slight headache and a general feeling of fragility. Then we went out to Cape Kidnappers, this time on the ground. Chris operates smoothly in the gannetry, simply walking between the nests, neatly tilting each bird to one side with a stick, to expose a leg and read the ring number. No birds fled in panic. It was all very orderly and satisfying. Try that in an Atlantic gannetry and they will go mad. The Plateau colony is marked out in sections using hose pipe and numbered wooden pegs – simple in a place so flat and even. Each time he read a ring he noted the nest contents, in effect compiling a dossier of individual life histories. It sounds simple, but this kind of information, so hard to come by, is vital for an understanding of population dynamics. The pity is that Chris is essentially a field naturalist and much of his hard won information still remains unanalysed. Others, with far less material, publish madly on the publish or perish principle.

## Dec 25th

A funny sort of Christmas Day. The twins were awake by 5 30, opening their very modest presents. Then we went a-wandering and found an interesting river by which to picnic. It was a vast spate bed with a fast, shallow flow. A pair of banded dotterel with their dull plumaged youngsters foraged on a gravel bank and reacted aggressively to a pair of completely harmless black-winged stilts. My friend John Busby would have made wonderful drawings.

## Dec 29th

The children loved Marineland, their first experience of a fairground. Usually such places seem pretty tawdry but here the twin-hulled boats on the little lake were quite charming. Really we went to take a gannet chick which had fallen off one of the Black Reef rocks. They are trying to build up a small breeding colony at Marineland and later I saw what I could never have imagined – full-winged adult gannets flying out from a captive colony to fish in Hawke's Bay and then returning to Marineland to feed their chicks. Absolutely amazing. What a nice way to breed seabirds in semi-captivity. It wouldn't work with our gannets at Edinburgh zoo. Can you imagine a chevron of gannets flying down Prince's Street? They might nest on Arthur's Seat.

Dec 30th

John Busby's drawings arrived for my book on seabird biology to be published by Hamlyns. They are absolutely splendid. There is nobody to touch him when it comes to drawing behaviour; others are woefully wooden by comparison. His semi-abstract oils are marvellously evocative, too. When you compare his art with unmade beds and pickled pigs, or was it sharks, you must admit that the one word 'art' simply cannot be stretched to cover both forms of expression; call a mucky bed what you like.

Jan 6th

We salvaged an old armchair and an ancient wooden settee with no padding from Ron's 'store'. They go well with our dilapidated cable drum.

Jan 9th

Twins turned up trumps today and I was proud of them. Ron took us to the Urewera National Park where his pal Lou Dolman is police officer. Lou is a legend in the force for his search and rescue work under hazardous conditions, not only for courage but also for endurance. A real Kiwi bush ranger, hill walker, handle-anything type, he has several citations for arresting armed offenders, including a murderer and has shot some 150 wild boar whose wicked tusks decorate his shed. His collection of old rifles and pistols has historical links to Maori chiefs and early European colonists. Sitting a trifle oddly with this rock-hard persona is a keen interest in ferns – but then he loves all nature. Lou, the genuine article through and through, seemed a bit contemptuous of Cape Kidnappers' tameness. He'd have visitors walking from Clifton carrying a half-hundred weight of rock apiece.

*The children and June with Lou Dolman & Ron Fisher*

We rattled around the Urewera Park for five long, hot and dusty hours in Lou's Toyota and this was after the children had endured a four-hour run from Napier. It was damnably hot, but to my surprise and relief they stuck it manfully – not a murmur. It really was far too much to ask of barely five year-olds. They didn't even have a drink until we stopped near a mountain stream where they knelt and lapped like dogs. Lou must have impressed them mightily. Maybe I should go out and rescue a few hikers; more likely I would lose myself!

Lake Waikaremoana is huge, limpid and blessed with delightful camping spots. The 'bush', mainly podocarps, gives way on higher ground to beech (Nothofagus) which resembles our birch though unlike Betula it is beautifully buttressed. Epiphytic ferns, mosses, lichens and Fuselia luxuriate everywhere and the forest ridges stretch away to the distance. This is the third largest National Park in New Zealand. The lake, one of the very few left in the country with a completely natural catchment area, is not enriched by fertilisers. Alas, the hydro board use it for generating and the resultant fluctuations in water level greatly affect the invertebrates in the lake and, through them, the fish. Biologists are trying to get the board to change the timing of their operations to less sensitive periods but so far without success.

Next day we were up in time to see Lou attempting to deal sudden death to an Australian magpie. They are a pest, destroying the nests of native birds which have already suffered severely as a result of alien species. Then we wandered up to high level Lake Waikareiti, through magnificent bush alive with tuis, riflemen, fantails and pigeons. Lou says we have now seen a bit of the real New Zealand. Pete, a biologist working on the lake, sent us home with a freshly caught brown trout of nearly 2kg – a real 'beaut' as they say.

Jan 13th

Back at the Cape we find the gannetry bursting at the seams with well-grown chicks, boldly black and white and far more attractive than the Atlantic gannet's mainly black juveniles, though we do have some lovely silvery ones. The pectoral band is most conspicuous, just as in many juvenile cormorants, a primitive trait possibly going way back to the gannet/cormorant ancestor.

A hawk moth chrysalis which had somehow found its way into our keeping hatched last night. I put the newly emerged moth onto the tree left over from Christmas where it clung to a cardboard ornament. By morning, another moth was on the window together with a lot of eggs. We liberated it into a eucalyptus tree but later found yet another in a corner of the room. The one we had liberated was a male that had come in response to pheromones (chemical stimuli) given off by 'our' moth, a female, which had then laid the eggs. The one we set free had the beautifully broad feathery antennae which the male uses to pick up the female's scent.

Jan 14th

The estuary near our Napier cottage was alive with birds. A kingfisher caught a crab and hammered it left and right on a post before swallowing it. A long line of shags adorned the old posts stretching across the estuary. The little pied shags have stubby bills; they eat a lot of crabs. The little black shags are pencil-slim with slender bills. These differences in related birds which share the same feeding areas help them to

*Australasian gannets and well-grown chicks*

divide up the food, one specialising in one item and another in something else. It is just the same with the Galapagos finches and is a general rule in nature.

I was given a good tip for sand fly repellent; mix dettol and an oil, say baby oil. I wonder if it would work with our Scottish midge.

Jan 24th
Ron took us to a fantastic museum near Bell's Clearing, packed with amazing relics, all assembled by one man and stuffed higgeldy-piggeldy into his farm buildings. The lovely old cars included a 1926 Rolls Royce silver ghost and a Buick. There were pumps, ploughs, kitchen ranges and utensils, washing tubs, flat irons, sewing machines, typewriters, watches, coins, medals and much else, mostly unlabelled. On one car the spare wheel was not a replacement but was meant to be bolted on to the side of the wheel that had a flat tyre.

Jan 27th
Another scorching day with a strong south-westerly. Everything is parched and brown, withered and burnt. How do sheep live on a few blades of dried grass? This month has seen four inches less rain than average and water is running low. What a change in the colony, too. Most of the young are fully-feathered and they wander around the colony in a way never seen in Atlantic gannets. Many chicks were way off their proper nest site, blundering here and there quite pointlessly. It certainly was not due to human interference. I cannot understand what advantage they gain. Even on top of Black Reef Rocks where nobody ever goes, there were some wanderers.

Jan 31st
June is descended from a long line of gypsies, or perhaps it is Bedouin, accustomed to carrying their house and belongings on their backs. We planned to set off nice and

early to Wellington where I am to give some lectures at the University but at midday she was still cleaning and packing. Piles of stuff lay everywhere. At long last the poor old Holden looked like a junk cart. She wanted to take the mattress and underlay from our double bed, together with four sleeping bags, pillows, tent, folding table, huge box of utensils, another of food, three cases of clothes, two haversacks, the children's rucksacks, water carrier, etc. To be fair, we were first going to camp in a National Park and explore South Island.

Almost a fortnight ago Hamlyns air-freighted galleys and photocopies of my forthcoming book on Seabird Biology to Napier airport. Alas we muddled up the telephone numbers of air cargo and the airport – two quite different establishments and thought we'd lost the lot. Maybe I should return to the simple life on the Bass Rock or, even better, Tower in the Galapagos.

*Wandering albatross*

Feb 1st

Thrilling views of wandering albatross as we crossed the strait between North and South Islands. What a wingspan. One can scarcely bear to think about the slaughter caused by long-line fishing in the southern oceans. The birds swallow the bait, hook and all, and drown. There are now ways of minimising this tragic damage but the problem is persuading fishing boats to co-operate. Cape pigeons, prions and short-tailed shearwaters were much in evidence. The charming port of Picton was loud with youths in souped-up cars and motor-bikes. Remember, you antiquated and anti-social bird-watcher, this is now the seventies, not the fifties. Fast forward to today and the seventies seem quite civilised, even old-fashioned.

Feb 2nd

En route to the Abel Tasman National Park. We were lucky to find a flat with all facilities for 12 dollars a night at Kaiterikiti. At last, a peaceful night. But just before midnight a squeal of tyres, slamming doors, loud music and shouting penetrated the wafer-thin walls. I stuck it until 1 0am (June sound asleep) and then went round. Blessed silence at last until Simon piped up "my insect bites are itching". No sooner coped with that than Becky was awake. Made breakfast at 6 30 and jollied the children along whilst mistress June slept soundly – or did she? I feel like a red-necked phalarope, where the male does all the domestic chores whilst the female wanders off and finds another male with whom to do the same thing.

*Weka stealing a sandal*

Feb 4th

What a tremendous climb to the campsite in Abel Tasman; hair-pin bends and murderous slopes, the bush dripping after heavy rain and the car totally steamed up. It was packed to the gunwhales. A major breakdown here would not be funny. We erected our borrowed tent as best we could, but by midnight the canvas began to flap ominously and then madly. Slamming blasts of wind billowed the walls and roof. Because of elastic guy rings and an external frame it was impossible to make the tent rigid and the poles dug deeply into the sand, loosening it even more. We were up well before dawn but things got even worse during the day so we shifted to a more sheltered site. Then after showing us what it could do, Tasman relented and superb weather followed, tempting us into the sea, and a marauding weka to steal my beautiful new leather sandals. At least I assumed it was a weka.

Feb 9th

On to Reefton, an open, airy, hill-cradled little town beside a wide sweeping river. Perhaps a bit shabby, but sedate with old-fashioned houses and sweeping gardens – I imagine a good place for a boy to grow up in before conquering the world. Then ugly Greymouth, the sand a dirty grey and the sea huge and turbid, followed by the climb between Franz Josef and Fox Glaciers magnificently precipitous amongst mist-shrouded mountains. The glaciers' dirt and 'rock flour' had turned the Fox river white. It began to rain blotting out the beautiful schist and granite in a grey pall like Scotland at its dreariest.

Feb 12th

We took the famous bridle path along the Skipper Gorge's dizzy, unfenced drops; hair pins, rough surfaces, shale edges – a real road. The Shotover river flowed fast, grim and turbid. I wouldn't fancy rough-water rafting down it, though people do.

Feb 21st

Walter Peak is 2000m and most of it is a sheep station. The house is open to visitors and we enjoyed a splendid tea in the large dining room with log fire, scones, savouries and cream pancakes. Then a demonstration of sheep dog work (a good dog fetches 2000 dollars) and wool spinning. Undeniably tourist stuff, but so well done that it felt like a visit to friends. The Station runs to 64,000 acres and raises highland cattle and sheep. The cattle, one of which came from Scotland's Braemar, are now worth 18,000 dollars a head, though don't ask me why they are so valuable. Old Mackenzie, who started the farm used to row 16 kilometres to Queenstown every week with fish and vegetables for market. It used to be a three and a half hour slog to the top of Walter peak; nowadays it takes seven minutes in a helicopter.

Land prices around Queenstown have rocketed and a small building plot in 1978 could cost up to 30,000 dollars. What price the millions of acres which early European settlers simply 'claimed' and then made out title deeds. No wonder indigenous peoples all over the world feel cheated.

Feb 23rd

There was a delay of one and a half hours at Lindis Pass. The gate is opened for a paltry five minutes every two hours and if you miss it - tough. Low cloud obscured Mount Cook so we didn't see the famous vista up lake Pukaka to the mountain. In Canterbury the friendly Wragge family welcomed us for the night; Graham is one of the part-time student rangers at Cape Kidnappers and seems destined for great things.

The Picton ferries were on one of their all too frequent strikes; the British disease has caught on. What is it about ferries that make them so strike prone? Think of the French and cross-channel ferries. We wasted a gloriously sunny day in a long queue and reached Wellington and the Wells' household at midnight. Rather than rouse them at that unsocial hour we all slept in the car. I imagine the children are not at all impressed by this trip but they won't remember it in any detail anyway. Then it was north to Napier and the ever hospitable Fishers before gladly turning into the familiar grassy forecourt of Cape Kidnapper's cottage the following day.

What a splendid place is Cape Kidnappers in which to wake up on a typically sunny New Zealand morning. The sky is clear blue, the little grey warbler singing, sheep and cattle grazing and on our very doorstep lies a bustling gannetry. What more could I ask? Sadly, though, this was to be our final stint at the Cape, so before leaving Napier I called at the offices of Lands and Survey to thank the generous Jack Campion and to present a copy of my newly published monograph on the Atlantic gannet. Much of the behaviour is very similar to that of the Australasian. Then it was off to Wellington and the noisy camp site where we were to stay. What price the peace of Kidnappers. Here it is sleep to the accompaniment of lorries, trains and pulsing jets of steam. Still, there are no possums dancing on the tin roof and it is very cheap.

After our peaceful life at Kidnappers I have what seems to me quite a load of lectures at Wellington University, but they are a friendly lot. The worst chore is getting into and out of Wellington itself. Parking at the University is unexpurgated misery and extricating oneself from the chaos at 5 0pm calls for a Houdini. Wellington has a lovely situation, but the city itself is ugly. I haven't seen a single really attractive street and the buildings are sheer boredom. But the excellent zoo has a specially good primate section (spider monkeys, capuchins, hamadryas baboons), a pair of magnificent Siberian tigers, and leopards, cheetah, puma and really healthy brown bears. Although in a way I like zoos, they do exhaust and depress me. Their primary use, it seems to me, is as gene banks. Extinction is for ever and a captive breeding population might just prevent it. And of course zoos are excellent for breeding animals that, although not in danger of total extinction as a species, may be locally rare. Then they can be released and protected, as has happened at Azraq.

March 17th

We have just returned from Palmerston and Massey University where I gave a couple of lectures. Brian Springett, professor of biology, hails from Durham University where he was a student of John Coulson, a great seabird man whose pioneering work on kittiwake breeding ecology broke new ground, showing how small differences in the position of a nest within the colony can significantly affect clutch-size and breeding success. Thirty years on, I am presently (in theory, at least) helping to supervise one of

Springett's Ph D students working on Australasian gannets. The Springetts are just re-covering from glandular fever. I hope we don't catch it. I'm a bit like the late David Lack, the erudite director of the Edward Grey Institute: he abominated being any-where near you if you had a cold.

I met dear old Kazimera Wodzicki and Sir Charles Fleming when I lectured to the Royal Society of New Zealand in Wellington. Kazimera was nearly 80 but still spry and enthusiastic and still involved with Cape Kidnapper gannets which had been his interest since the 1940s, along with Chris Robertson senior. Like many older people, including me, Kazimera, an aristocratic Pole of great courtesy, was not exactly terse, but I found it highly discomfiting during question time when some pompous idiot took it upon himself to interrupt him, telling him to come to the point and stop hog-ging the floor. At that time, Kazimera was supervising Elspeth Waghorn working on the foraging habits and growth rates of gannets.

March 21st

Left for Christchurch University on a filthy grey morning and arrived in torrential rain. There was no sign of the diminutive John Wareham who was supposed to meet me. John, a vastly experienced seabird man, originally from Yorkshire, was instead sipping tea in the zoology department under the impression that the airport was closed. I think he might at least have checked. That laid back New Zealand philoso-phy 'she'll be right' may be admirable but… !

I chickened out of a scheduled visit to Dunedin University in case I got stuck and in the event I was wise. When I got back, Wellington was in bright sunshine but the mist rolled in and by 4pm the airport was closed and remained so for the next two days.

Before the long haul back to Aberdeen I had to dispose of the ancient Holden which had served us so well. None of the local garages was interested so Don Merton generously offered to shoulder the thankless chore of selling it for us. I hope it isn't an insult to a Kiwi to say this but Don is a real gentleman. Later he became Government Conservator on Christmas Island.

The teaching stint in Wellington closed our hugely enjoyable six months in New Zealand and after a monumentally boring flight (33 hours all told) we were met at Heathrow by Trevor Poyser, another of Nature's gentlemen. After that it was just a matter of picking up the old Volvo from the barn in which it had languished all winter (true to form it started first time) and motoring north through filthy English weather. After the brilliant sunshine of Hawke's Bay the dreary midlands seemed doused in grey. The Scottish borders came as a great uplift to the spirits and we holed in at Thorlieshope Farm for the night. Oh no! New Zealand mutton for supper.

The trip ended on March 31st in Aberdeenshire's Buchan on a fine late winter's day with washed-out pastel shades and the curlew's liquid trilling everywhere. Back at our croft we turned on the water supply and it poured through the living room ceiling from ten bursts. Welcome home.

*Sealion juggling*

*Black-winged stilts*

*Fantails*

*Tuis*

*Gannets display*

## Chapter 13

# LESSONS FROM CAPTIVES

*"There is an amusing belief among country boys that an owl has to turn its head to watch you and must watch you if you are near him, so that if you will only walk completely around him he will wring his own neck"*

A.W. Caldwell & G.E. Lumsden  1934 'Do you believe it'

August is a dismal month for Ailsa's gannets, much worse than for Bass birds. At this time of year scores of doomed youngsters sit stoically amongst the nettles and rank grasses which festoon the base of the cliffs, fouled by droppings and moulted feathers. Some have overbalanced and fallen from their ledges, perhaps whilst exercising their wings, at this stage heavy with blood-filled quills. Or maybe they fell whilst defaecating over the edge of the nest; they are not the most agile of birds. But many are knocked off their pedestals by adults landing or departing clumsily. At a certain stage in their growth, when heavy and inept, gannet chicks are particularly vulnerable to dislodgment. The base of the Bass cliffs is washed by strong tides and fallen chicks swiftly drift away, often attended by opportunistic great black-backed gulls tugging, pretty ineffectually, at the vent until they tear it enough to really get in. It is difficult for them to get a purchase in the water despite their determined back-peddling. The base of Ailsa Craig, though, is fringed by slopes and boulders. My very first visit, now more than fifty years ago, had been a character-forming few days, ringing gannet chicks during one of those hot, energy-sapping spells which can transform the traditionally mild Ayrshire coast from a dispiriting greyness into a passable imitation of the sub-tropics. My friend John Leedal and I were picking our way beneath the cliffs when we came across three large, undamaged gannet chicks squatting forlornly in the rank vegetation.  If only we hadn't found them! But we had, and it was unthinkable to kill them, nor could we just leave them to starve to death or be eaten alive by the rats which at that time infested Ailsa. The only solution seemed to be to put them into empty gannet nests and just hope that the adults would foster them. Jimmy Girvan, the quarryman who at that time lived on Ailsa told us from long experience that such substitution did not work but what else could we do? Years later, on the Bass Rock I found that things were not as clear-cut as Jimmy had thought. Whether a substitution would be accepted depended on quite a few things. If the putative foster-parents had themselves recently lost their own chick they would accept a substitute, even of a different age than the one they had lost. But in a few instances in which we had unwittingly chosen unsuitable foster-parents they attacked the orphan mercilessly and either forced it off the nest or killed it. So, with palpable ill-grace, we picked up our fallen Ailsa chicks. They weighed perhaps 2,000g apiece and used their beaks wherever they could. A heavy, fluffy gannet chick tucked under each arm with their bills claiming the full attention of both hands rapidly led not only to aching muscles but to an overwhelming desire to scratch innumerable maddening tickles. Carrying them over slippery boulders didn't help, but we struggled on and up to the cliff-top nests where we dumped them in a nest a-piece and fled. At least they had a slim chance of being adopted whereas it would have been slow and certain death if we had left them at the base of the cliffs.

Twenty years later I came across two more chicks. This time I took them back to Aberdeenshire with romantic notions of rearing them and perhaps even persuading them to breed like E.T. Booth had done in the late nineteenth century. It proved a costly mistake in every way, though I did learn a few things. Really, with very few exceptions, the only acceptable way in which to keep a wild creature captive is under passably natural conditions, hardly possible for a gannet, though Napier zoo managed it. Gilbert, a male, was half-grown whilst J., a 10-11 week-old female, had acquired all her feathers but traces of down remained on her head. Both would inevitably have fallen prey to rats. On some Pacific islands the Polynesian rats literally gnaw their way through the backs of incubating albatrosses apparently without the birds doing anything to stop them. The sheer ineptness of seabirds in responding to predators to which they have not evolved an instinctive reaction is truly astonishing. Masked boobies in the Galapagos simply sit there whilst the small black Darwin's finches peck the base of a wing quill and then sip the blood. I have seen blue-footed boobies stand around fidgeting whilst mocking birds pecked their cloacas raw and then drank the blood. Albatrosses and boobies may not be the brightest stars in the ornithological constellation but one might still imagine that a little academic curiosity about the source of the pain would not have been beyond them.

J., Gilbert and an injured adult which we had picked up soon accepted piles of broken concrete slabs, not quite like the lovely perpendicular columns of green, veined granite which they had vacated on Ailsa, but no slab of rock anywhere can be the last word in comfort. Although I knew I faced a hard time feeding three gannets I hadn't really considered, soberly, the incredible amount of fish they would devour nor the problem of keeping it fresh. Obviously I had to lay in bulk supplies and on many a bitter Aberdeenshire winter morning, before setting off for work, I cursed my sheer lack of commonsense as I hacked frozen slabs of flatfish apart and tried to thaw it enough to tempt those grim sentinels sitting impassively on their icy concrete piles through hail and rain, wind and shine. I felt horribly guilty when I saw them stoically sitting out another short, bitter winter's day before facing an interminable night. I felt it would have been far better to have let them die on Ailsa. But then, as they not only lived but seemed healthy, I began to dream that in time they could be released with some chance, albeit slim, of survival. I tried keeping them in the stable but they seemed happier, if that is the appropriate word, outside. Indeed gannets can cope with the worst that the North Sea and North Atlantic can throw at them in winter and not much can be colder and wetter than that.

*Our captive gannets*

By the age of ten months they had acquired adult voices in place of the juvenile 'yap' and their brown irises had lightened in colour though not yet the cold grey-blue of the adult's. I was fascinated to see how their behaviour would change with age, which I could never have observed in the wild. Birds inherit not only the physical equipment, nerves, muscles and plumage necessary for the behaviour such as the various displays used in communication but they inherit, also, the motivation which compels them to act when the appropriate stimuli, such as a receptive female, come along. Much of their behaviour does not have to be acquired by watching and imitating other members of their own species; it is already 'wired in'. The so-called 'blank slate' psychologists were quite, quite wrong when they believed that it all has to be learned. So it was hugely compelling to see which of the genetically programmed behaviour patterns would emerge and how complete or otherwise they would be.

Even if they did perform some of the gannet's ritualised displays it would not prove beyond doubt that there could not have been any learning involved; the purist could always point out that, whilst in the nest, they would have seen adults displaying. But at the very least it would prove that they were astonishingly prone to acquire their own species' display long before they could possibly use it. They did not acquire any herring gull display, though they had seen plenty of that too.

*Great skua pulling a gannets tail*

In the wild, during its time in the nest, the young gannet does not usually perform any of the adult's territorial or sexual displays. In this it differs from its cousins the boobies, several of which whilst still chicks produce perfectly recognisable versions of these displays. It might well be that there has been strong selection pressure against gannet chicks doing so because such displays could elicit strong reaction from an intruding adult which, in the case of the cliff-nesting gannet would be desperately dangerous. The chick could so easily be dislodged with fatal consequences. From the beginning, though, Gilbert and J. performed all the usual non-display behaviour such as preening, begging, scratching, stretching and juggling with sticks and moulted feathers. Juggling is useful practice for handling and swallowing fish.

At about 39 weeks Gilbert started to perform the gannet's territorial display, at which point his behaviour became extremely interesting. Basically this display is ritualised ground-biting which gannets employ to warn off potential intruders. At the very time that Gilbert began to perform this site-ownership behaviour he approached me and attacked my shoe, subsequently mounting his 'nest' (the cement slab) and performing the polished version of the gannet's site-ownership display. The female, J., even though five weeks older, did not develop territorial behaviour at this time. In the wild, females do not establish the site.

But the most interesting behaviour shown by Gilbert proved completely new to me. I have watched gannets for thousands of hours but I had never seen one behave quite as he did. And he repeated it several times. The behaviour itself was not particu-

*Gilbert bowing*

larly spectacular, merely a spasmodic jerking of the head with the neck retracted, head thrown back and bill pointing upwards, but it was hugely intriguing for two very good reasons. First, it demonstrated that a complex behaviour pattern, a potential display, could lie bright, shiny and unused in the gannet's neuromuscular toolbox. One tends to forget, or at least I do, that the lovely ritualised displays which so many birds possess, are not the only ones that they could have had, or could still have in the evolutionary long-term, but merely a selection, brought about by goodness knows what complicated and possibly fortuitous combination of evolutionary events. Of course some behavioural building blocks will be used in certain circumstances rather than in others. For instance the false-drinking and associated head-dipping found in many ducks during courtship is facilitated by their aquatic habitat . But Gilbert showed me that gannets have complex, apparently stereotyped behaviours (displays) 'ready-made' but not used in their normal repertoire; a revelation with quite some implications though I'm not sure what.

Fascinatingly, Gilbert's display quite uncannily resembled the sexual advertising display of some of the cormorant species. Gannets and cormorants, though fairly closely related, have been distinct for millions of years. Yet this bizarre combination of movements still persists as part of the gannet's inherited capability even though rarely if ever used. The likelihood of Gilbert's display being merely a chance combination of movements that just happened to mirror the cormorant's display is vanishingly slight.

Thus the cormorant's display, perfect in form, still exists within the gannet's genome alongside the 'new' display that the gannet now uses in a sexual context; the 'new' display looks nothing like the cormorant's head-jerking. This makes complete sense when one knows that current gannet sexual advertising in fact derives not from the ancestral gannet/cormorant advertising display but is a completely separate display. There seems no reason why the old, now superceded one should not still exist as it were 'buried' alongside the new one. Why the phylogenetically more primitive cormorant/gannet display became suppressed in the gannet is, like so much in animal behaviour, unknowable.

Alas, Gilbert died whilst still immature and J. and the adult went to Aberdeen zoo. Gannets will breed in captivity. In Napier zoo the Australasian gannet breeds quite successfully. Given an adequate environment, large enough for them to fly and dive (indeed a tall order) they would be fascinating birds to keep but without an adequate environment they are pathetic captives, painful even to see.

Rearing fallen gannet chicks for eventual release is a totally different matter. Every year Fred (who died in 2008), Chris and Pat Marr rescued several from the base of the Bass Rock and fed them before eventually releasing them out at sea. Their behaviour shows, unmistakably, when they are ready to go. Before that magic moment they are fairly inactive except for preening and wing-exercising but when the decisive moment approaches they become desperate to get to sea. At any time from late August on, this motley collection of unfortunates hung around the Marr's back garden, some still with lots of down, others in the black plumage of the juvenile. At feeding time some gently took fish from the hand like well-bred dogs, whilst others remained violently aggressive, never failing to bite the hand that fed them. I found exactly the same thing when ringing gannet chicks. Some were conveniently docile whilst others went beserk. I suspected that the more aggressive chicks were males but no doubt individual differences occur in gannets and other animals as well as in us. Apologists for the idea that humans are born with 'blank slate' minds, ready for environmental experiences to write on, rather than with inborn traits, surely cannot be biologists. But an interesting aspect of the readiness with which the human mind will embrace even the most bizarre ideas must be that it is one aspect of the open-mindedness which has proved to be so spectacularly successful throughout our evolutionary history. We readily try anything and believe anything, no matter how bizarre or even impossible. All too often this proves a tragic trait, allied to one form or another of religious belief.

*Mockingbird sipping blood from a blue-footed booby cloaca*

*Laysan albatross feeding young and on the sea*

Chapter 14

# THE P(F)LIGHT OF THE
# PLASTIC ALBATROSS

*" I once found a list of diseases as yet unclassified by medical science and
among these there occurred the word 'Islomania' which was described as a
rare but by no means unknown affliction of the spirit ….
an indescribable intoxication"*

R.J. Berry 2009 'Islands' Collins  London

How did I come to be suspended over this boiling maelstrom between Kilauea Point
and a tiny offshore rock called Moku-ae-ae, in a hired helicopter carrying a load of
life-sized plastic albatrosses?

In the beginning was the gannet. Then the boobies – all six species. I owe a lot to
boobies. Indeed my life has been gannet-booby driven.  Boobies opened the door to a
Conference of the Pacific Seabird group in Hawaii, to which I was an invited speaker.
I delivered quite the worst talk I have ever given and all because of a misunderstand-
ing. I thought my talk was to be about gannets and boobies, which would have been
easy, but to my consternation, when I got there, I was casually informed that they
wanted a light-hearted, amusing, after-dinner speech of about 20 minutes. If there is
one kind of speechifying that gives me palpitations it is after-dinner; easier to relax in
the dentist's chair. Put me in front of The Royal Society, stuffed with their eminences,
and let me talk about brood-size and reproductive strategy in boobies and time flies –
at least for me. But to amuse a gathering of distinguished Americans? Well, as John
McEnroe said, "you cannot be serious".  But Americans are very kind.

Richard Podolski, one of the participants and famous for re-introducing seabirds
to islands from which they had been driven by pests either human or introduced,
thought I might like to take part in a small expedition to plant plastic albatrosses on
Moku-ae-ae. If the plastic models succeeded in attracting real Laysan albatrosses to
this little rock they would be totally safe from rats, cats, dogs and humans. And the
models themselves would be secure from interference. The idea that they might attract

the real thing was not entirely fanciful either. Richard specialised in this esoteric art and had already successfully introduced puffins to a former breeding island from which they had disappeared. He and his team holed up on Kaui near to the ancient lighthouse on Kilauea Point busily turning out dummy albatrosses, no bathtub play-things but life-sized Laysan albatrosses in a variety of authentic display postures. They made them by the dozen, all nicely painted and with stiff wire legs that you could stick firmly into the ground. They dotted these plastic goony birds amongst the trees which surrounded the biologists' work-shop-cum-living quarters, at intervals the models emitted the weird whoops and bill-clattering which real courting albatrosses do. The vocalisations came from speakers artfully embedded near to the models and fed by tape-recordings. They made a brave sight, statuesque black-and-white models gleaming in the sun, but alas they could not move or lay eggs. Real albatrosses lived not too far away, for they nest by the thousand on Midway Atoll, but attracting them to nest on a new island seemed a daunting challenge. Seabirds can and do extend their range if conditions such as food and safe nesting sites are suitable. Usually (though not inevitably) this is achieved by a few pioneers colonising a new breeding site and then attracting others. Sometimes the new colony grows rapidly but this depends on a number of things, such as the nesting habitat being really suitable and there being a plentiful source of new recruits. Our putative colony of albatrosses on Moku-ae-ae would never be other than modest, if only because space was so limited there, but it could act as a springboard.

The art of introducing or re-introducing birds to suitable breeding locations has become increasingly important as more and more species become threatened with

*The old French lighthouse on Kilanea head, Kauai, Hawaii, and the cottage where plastic Laysan albatrosses were made for transplanting on the islet of Moku-ai-ai*

*Plastic Laysan albatrosses on the Hawaiijan Island of Kaui at Kilauea Point. These were reallistically spaced in display postures to attract prospectors from (probably) Midway Island where there is a large colony*

local or even total extinction. Some rescue schemes such as the re-introduction of the Californian condor, the sea eagle in Scotland or the red kite in Britain are hugely expensive, involving captive birds which even after they have been released into the wild still have to be fed for a long time before they are able to fend for themselves. Hunting skills can take a long time to acquire, though I don't understand why some skills seem to take so much longer to attain than others. The peregrine falcon's stoop must surely rank as an outstanding feat of physical coordination, yet this doughty falcon can breed in its second year of life whilst many seabirds require four, five or even many more pre-breeding years. Size probably comes into the equation as well, though certainly not in a simple way. However I'm not going to dip my toe into that particular water. With seabirds it is usually a case of attracting wild birds to suitable islands or providing better conditions and safeguards for the remaining survivors of a once-successful population.

In the case of the Laysan albatross of Kilauea point and Moku-ae-ae it was a matter of tricking them into settling on Kilauea and subsequently attracting them to this nearby rock on which, in all probability, they have never bred. And trick them Richard did. Dotted around the peninsula in open spaces between the shrubs were entire groups of model albatrosses, their groupings and display postures cleverly simulating a breeding colony in full swing. The "moo" calls and 'bill-clunks' coming from appropriately sited loud-speakers had been recorded in the teeming colony on Midway Atoll. Everything in the albatross's considerable repertoire of calls and postures was also thrown in and the whole set-up must have hit overflying albatrosses like a burst of flak. Prospecting albatrosses are likely to be young birds, perhaps between three and five years of age, not ready to breed immediately but interested in establishing themselves in a suitable locality. An area which already contains displaying birds is much more attractive to such prospectors than an empty space, or even than an area containing mainly long-established pairs. The pattern in which the models were set out was artfully designed to be about half as dense as that of a fully established colony so that potential incomers would see the space. Many seabirds do seem to be attracted most strongly to a colony which, although well-established, still has room within it. A salutary lesson to that effect was provided by a Scottish island (the Isle of May) which had a huge colony of herring gulls to the detriment of other seabirds. They were heavily culled but, alas, the spaces thereby opened up attracted large numbers of new recruits with the result that the colony actually increased. In effect the established birds sent out the message "it is good here" and the spare space added "come and join us," which they duly did.

Richard wanted to discover whether the albatrosses vocalisations played a part in attracting over-flying prospectors. One would naturally think so but to demonstrate it he gave one group of dummy birds loud-speakers which emitted albatross courtship noises whilst the other group was silent. As a control he left some areas completely empty and silent; all they had to offer was available space. After setting all this up it was merely a matter of sitting at the top of the old French lighthouse on Kilauea Head with a telescope, waiting for albatrosses to fly over the colony and then recording their reactions. Would they be interested enough to land? If they did, would they approach one of the models and react to it? Or would they treat the whole show with complete indifference? The results were interesting. Where there were neither models nor vocalisations hardly any albatrosses landed. Where there were only silent models about 5% out of 1,300 instances of overflying albatrosses landed whilst models coupled with calls attracted more than 8% of 1,000 overfliers. Whilst not wildly spectacular these differences did support the commonsense supposition that both sound and sight are important. Also in line with expectation, models in display postures were more attractive to overfliers than birds which were merely resting. Perhaps more surprising, pairs attracted more birds to land nearby than did singles. Presumably members of pairs – and this includes us – do send out the message "I am attractive enough to have secured a mate" which unattached males may take as an invitation. In humans the long trail of disrupted unions demonstrate that single males indeed do just that.

In albatrosses it may take several years before enough pairs aggregate to create the social atmosphere of a 'colony', which is necessary for successful breeding. But bearing in mind that there are at least 30 species of colonial seabirds which are globally threatened in one way or another it seems an excellent idea to lure them to safer locations, especially when their predicament is due to introduced predators. So Richard's work is of great practical importance.

Our helicopter flight to Moku-ae-ae came after a stormy night which made a boat passage totally impossible even if there had been anywhere to land. The lava cliffs

*Planting plastic albatrosses in display postures, on the inaccessible rock called Moku-æ-æ*

were smothered in foam, the sea a boiling cauldron. Misty curtains of spray drifted high above the cliffs. On Moku-ae-ae itself there was not even enough level ground for the helicopter to plant both skids and so the pilot had to balance on one whilst we all climbed out clutching our plastic albatrosses and ducking nervously under the rotor blades. The fun part was setting them up here and there in display postures to give the appearance of an active colony in full swing, though alas without the noise. This kind of practical conservation work was quite new to me and seemed to be a splendid way to use the knowledge and insight gained from the countless hours of observing and interpreting behaviour. One could easily have imagined, wrongly, that such studies would have no practical value.

Another conservation project on Moku-ae-ae centred on Hutton's shearwater. Unlike the albatross, shearwaters already nested there but were being provided with artificial nesting burrows in the form of sections of conduit piping, some already being used. These days seabirds all over the world are under enormous pressure. The headlines rightly emphasise the extreme cases like the 100,000 albatrosses being killed every year by long-line fishing but the fact is that the great majority of the world's seabird species, especially in the tropics and sub-tropics, have declined hugely in the last fifty years. Erstwhile teeming colonies have dwindled to comparative insignificance and many have disappeared altogether. The wearisome litany of woes scarcely needs to be repeated – overfishing, pollution, introduced rats, cats, dogs, pigs and goats and of course direct and continuing plundering of breeding colonies by humans. So it is hats-off to those who are trying to get rid of artificial predators from islands and to protect seabirds from depredation by man. The two things of vital importance to seabirds are good fish-stocks and protected breeding colonies. To provide them means changing human nature. We need battalions of Richard Podolskis.

*The conduit pipe here is for the Hutton's shearwater to nest in*

*Laysan albatross display*

*Greenland white-fronted geese*

Chapter 15

# BONNIE GALLOWAY AND MAGICAL SEABIRD ISLANDS

*"Grey, gleaming Galloway".*

*" Blows the wind today and the sun and rain are flying,*
*Blows the wind on the moors today and now,*
*About the graves of the martyrs, the whaups are crying.*

*" So some love Merle and Mavis,*
*and say, in a garden close,*
*That Philomel most brave is*
*Who sings to the darkling rose:*
*But others love a bill able*
*To spill wild music through*
*Sweet syllable on syllable*
*And cry, Curlew, Curlew".*

The soft Galloway rain drifts over Bengairn. Who could believe that this secluded corner of rural SW Scotland was once a rumbustious mining community? We live a little over a kilometre inland of Balcary Bay guarded by whale-backed Hestan Island, the long-ago retreat of the Scottish king Edward Balliol of Sweetheart Abbey and of Oxford's famous college. In the woods below our cottage lurk deep holes from which came rock, rich with shiny black nodules of iron-ore. It was horse carted down to Balcary Bay and shipped across the Solway to Workington on the English side. Railway lines forged from that ore ended up in India or Peru for all I know. In 1964, on our dizzy descent from the high Andean village of Huancayo to the Guayas plain, thousands of metres below, we may have travelled on rails forged from our ore!. These days tawny owls hoot in the woods which cover the scars, now misted with bluebells in Spring. Blackcaps, willow warblers and, in lucky years, pied flycatchers and red-starts nest here. Stockdoves grunt and cushats "roo-roo" the summer days away. Have you noticed that woodpigeons usually say 'roo.roo/roo.roo/roo.roo/roo? It took me years.

Poor old Auchencairn. The village once boasted two inns and two churches, a school, a blacksmith, several shops, a muddy main street and even a thatched cottage or two. A garage has replaced the blacksmith, and with the inn, the school and the post office, constitutes the heart of the place. In Robert Burn's Galloway smuggling was endemic. French brandy came to the cave-fretted coastline via Ireland and continued by packhorse to Edinburgh – almost openly, it seems, for the smugglers were so numerous, organised and well-armed.

After World War II Auchencairn narrowly escaped total dereliction, for the rapacious MOD had its predatory eye on it as a firing range. In this crazy world beauty and wildlife play fifth fiddle to politics and the military and in sparsely-populated Galloway, well off the beaten track, a firing range debouching onto the Solway Firth, an ideal mud-basket for the uranium-depleted shells that over-shot the range boundary was just the job for the MOD. Instead, partly due to the influence of the MP for the area, who happened to live in Auchencairn, the range-fate befell 5,000 lovely acres near Dundrennan Abbey. It swallowed up farms, cottages, an old flour-mill with a water-wheel, an ancient graveyard and a delightful old walled garden which all fell into ruin. But, as on many MOD ranges, wildlife flourished. Ravens and peregrines nest on the sea cliffs, well over 200 badgers breed, roe-deer, almost too plentiful, are culled each year whilst barn-owls ghost along the hedges and regurgitate their black pellets of bone and fur amidst the high-voltage insulators in spooky MOD installations. And there is now a large and flourishing population of red kites in Galloway

Kite

Much of Galloway is marginal land, third rate for agriculture, full of boggy corners thick with meadow sweet and great for birds. The endless kilometres of stane dykes are shaggy with moss. When, late in the day, the winter sun throws low, slanting rays the shadows outline ancient strip pastures, the ghosts of an old farming method for impoverished land. Here and there the rocky coast is broken by tiny sandy beaches. On our side of the Solway the cliffs, though some distance from the open sea and not spectacularly high, still attract guillemots, kittiwakes, razorbills and fulmars, true seabirds one and all, unlike the scavenging herring gull, cold-eyed habitue of rubbish dumps. The Solway is a muddy old Firth, mostly grey and turbid with wave-stirred sediment. But the cormorants like it and their white-washed colonies stain the cliffs at intervals, though the ocean-loving shag avoids it. Early in March when snell winds still finger the dark, soddened ledges, great black cormorants resplendent in fresh breeding plumage, with frosted head-plumes and bold white thigh patches gather to breed on favoured stretches of cliff or on rocky outcrops barely above the

250

*A cormorant colony at Portling, Galloway, note the widespacing of nests compared with a gannetry*

*Cormorant colony with well-grown young, PortLing, Galloway*

splash zone, only to then forsake the site for a year or two, or maybe more, before returning. Often their going has nothing to do with disturbance or with breeding failure, although it is certainly true that cormorant colonies sometimes do move because of persecution. Seabirds vary in their faithfulness to a particular breeding locality. At one extreme gannets will put up with centuries of gross disturbance, including annual massacres as on Sule Sgeir and formerly on the Bass and Ailsa Craig, and still remain faithful, whilst at the other extreme tern colonies will up and away at the drop of a hat.

Cormorant courtship is bizarre. The male flicks the tips of his tightly closed wings upwards, which hypnotically covers and uncovers his two big white thigh patches. They blink like monstrous eyes whilst the female looks on, long-necked but, often enough, is not impressed enough to stay. I would love to know why. How does she choose?

Out on Scar Rocks in Luce Bay, well to the north of the Solway, gannets have built up a thriving colony of more than 2,000 pairs although they will soon run out of space. More than likely the pioneers came from Ailsa Craig, a few kilometres to the north. I have landed on the Scars only once for it is a tricky business with slimy ledges and even in fine weather a nasty swell. Strong currents boil around the Rock, sluicing wickedly past the menacing fangs which protrude here and there. You have to take a rubber dinghy close in, and it calls for nice judgment to scramble onto a ledge at just the right instant. How the Marrs have, over the decades, managed to land thousands of visitors on the Bass without ditching anybody is a miracle to rival the

*Cormorant display*

*Scar Rocks, Luce Bay, south of Ailsa Craig (75km) may have held breeding gannets by 1883 but post 1939 it drew in recruits and now holds more than 2000 pairs and is nearly "full"*

leaning Tower of Pisa, which some of them closely resemble. The Marrs used to be official boatmen for the Bass. Now the Scottish Seabird Centre in North Berwick runs landing parties.

In winter the Solway Firth attracts thousands of sea-ducks, grebes and divers. Identifying them far out on a choppy sea against a sullen grey sky can be a real tease. But above all the Solway is goose country. Peter Scott loved it and was the first to identify the Greenland race of the lesser white-front, a few hundred of which winter around Loch Ken each year. But the real pulse quickener is the vast flock of barnacle geese, some 14,000 of which, each autumn, beat a yearly passage from the wastes of Spitzbergen to the Solway. Their sky-obliterating flocks rise from the merse in a crazy bedlam of yapping. The big mystery, though, is the disappearance of the Bean geese, once common around Loch Ken but never seen nowadays. Nobody knows why.

Solway mud! Could anything be more stinkingly glutinous, more hell-bent on sucking your boots off and capsizing you? But it is rich in lugworms, cockles and trillions of small molluscs, a banquet which attracts tens of thousands of waders. At the edge of the tide masses of piping oystercatchers probe for cockles, but in recent years cockles have been hoovered up by the million using suction dredgers. And they have been gathered in thousands of sacksful by cocklers on foot, many of them from distant city suburbs, especially Newcastle. For years tractors raked them up, destroying untold millions in the process. It was a weird, eerie and dangerous business far out on the treacherous estuary in the pale shifting moonlight, or in the grey dawn, ever on the watch for the incoming tide. Solway tides are fearsomely fast and swallowed up more than one mired tractor whilst its driver hastily escaped on foot, lucky to get away. But to say they can outpace a galloping horse must be an exaggeration.

*Barnacle geese*

Galloway though bonny is more than a trifle damp – just look at the moss-draped dry-stane dykes. But it is scenically varied, beautiful and uncrowded which seemed to us a good reason to move there from our Buchan croft. At the time of our house-hunting we were fostering a fledgling carrion crow which Becky, our daughter, had charmed me into stealing from a nest in our garden. This corbie was now half-feathered, ugly as sin, touchingly confiding and insatiable. Normally in the nest it defaecates over the rim but with us it lived in a cardboard box on top of our dining-room storage heater. We couldn't leave it behind and our nearest neighbour, the friendly 80 year-old widow of the local shepherd, could never have been persuaded to feed a corbie. In any case she was too lame and the thought of her cosy little cottage harbouring a lime-squirting crow was too ludicrous even for us. So we kept it in its box and fed it on fragments of road-killed rabbits which, after due metabolic process, eventually whitewashed the interior of our old Volvo. In this malodorous condition we approached the old manse of Balmaghie church on a fine July evening. A blackbird fluted languidly from a mature chestnut tree and serene Loch Ken formed a perfect back-drop. Our visit, completely unplanned, was the fortuitous result of the 'For Sale' notice at the roadside but – luck again- Douglas Hutton, handsome, urbane and impeccable, answered the doorbell. From its perch on his elegant shoulder an imperious Amazonian grey parrot surveyed us keenly. We, with our scruffy young crow and Douglas with his beautiful parrot quickly established a rapport and agreement on the sale of the old manse occurred almost as an afterthought. So we swapped our

*Ducks and waders
at Mereshead
Nature Reserve,
Dumfries*

Aberdeenshire croft for an impressive old manse with stables and a wine-cellar. We lost our granite-dust tennis-court but gained a croquet-lawn. Of the original 18th century building only the cellar remained, together with the old stables, the iron rings still bolted into the walls for the horses of parishioners from over the hill at Laurieston. The cellar of the old manse, at the bottom of steep stone steps hollowed by centuries of feet and ideal for storing home-made wine, soon began to fill with dandelion-flower, sweet and subtle, and not-so-subtle rosehip, the colour of strong urine. Gorse-flower wine had exacted a blood price.

The real jewel, though, was the 20 kilometre long Loch Ken, enchantingly visible from our upstairs loo. A few hundred metres from the manse Loch Ken's fringes attracted hundreds of nesting black-headed gulls. Soon after the first snowdrops had bent their heads we would wake up one morning to their raucous but oddly appealing clamour, as welcome as the cuckoo in spring. They are really marsh gulls, more tern-like than the cold-eyed herring gull or the thuggish great black-backed. Soon their nests, built low in the marsh, held the green or khaki eggs, blotched with black, which used to be sold in London markets as plovers' eggs, and doubtless tasted just as good. Alas, many of their nests were washed out by sudden surges released from reservoirs above New Galloway. The water flowed into Loch Ken, raising the level by the metre or so which did the damage, before flowing over Glenlochar weir and then downriver to the hydro-electric power station on the River Dee at Kirkcudbright. But

*Loch Ken - 11 miles long and a favoured haunt of the Greenland white-front and other geese. Peter Scott first discovered them. They are now in decline*

within a few years the willow-scrub spread into the gullery and overwhelmed it. The gulls have now gone.

Far to the north of Galloway lie the fine wildernesses of Scotland's wonderful western and northern coasts. The sea is studded with austerely beautiful islands, romantic, impoverished, and at times tragic. St. Kilda, North Rona, Canna, Moussa, Rhum, Sule Sgeir, Sule Stack, Sule Skerry, Mull, Iona, Islay, Fetlar, Fair Isle and Foula. Although a mere Sassenach I have set foot on all of these except the three 'Sules', and even these I have sailed closely around. In 1998, quite out of the blue, 'Noble Caledonia' invited me to accompany two of their 'Seabird Islands' trips. For my supper I had to give a few illustrated talks on seabirds and lead the expeditions ashore if anybody wanted to be 'led', for they were an independent lot, mostly from pretty high-powered jobs. The ships, 'Professor Molchanov' and 'Professor Multanovsky' were smallish, Russian ex-survey vessels, which may have been used for 'surveillance' work, and therefore were not too luxurious. A second advantage was the small number of passengers – on the second cruise fewer than 30. We always anchored well off-shore and landed on the islands by zodiac rubber dinghies. It was all very sociable, unregimented and far from a glossy-brochure affair.

St. Kilda, 160 kilometres west of the mainland and, except for Rockall, the westernmost part of Scotland, was an old acquaintance. I first dropped anchor in Village Bay on Hirta in 1982, with Morton Boyd, at that time a Regional Officer in the (then) Scottish Nature Conservancy. We sailed from Stornoway in an old, flat-bottomed army landing barge which reminded me of the asthmatic 'Cristobal Carrier', also ex-war department, which took me and June from Guayaquil to the Galapagos in 1963.

During the rough crossing we sheltered for a while off Sanday. Under the circumstances the cook's offering of mince, rhubarb and bilious-looking custard seemed less than inspired. Morton was particularly interested in the centuries-old population of Soay sheep on Hirta; hardy beasts, a bit like the sweepings from a shearing shed stuck onto four feet and a head.

St. Kilda's gannetry on three precipitous stacs – Boreray, Stac Lee and Stac an Armin – is the world's largest although the Bass Rock now holds more than any one of St. Kilda's three. Stac an Armin is the highest stack in Britain though I'm not sure exactly how to distinguish a Stac from any other relatively small, rocky island. On St. Kilda in 1844 a great auk was the last British record before extinction. For sheer ro-

*Gannets around the Stacs*

*The gannet Stacs (Boreray, Stac Lee and Stac an Armin) from the main island*

manticism few if any of the world's remote islands can compare with St. Kilda. For centuries the amazing St. Kildans clambered around its awesome precipices against which the gannets dwindle to dust-motes. The nesting mass covers the rock like white icing, lit by shafts of sunshine which set them brilliantly against the dark rock and sea. Morton, a good cragsman, has climbed amongst them like the St. Kildans of old and it was Morton who had the first good shot at estimating their numbers – no easy task. A large gannetry cannot be counted with better than about 90% accuracy and you cannot achieve even that by a direct count from the sea. I think anybody who has tried to count dense masses of gannets, one rank partly obscuring those behind, from a tossing boat – or even a still one – will agree with that. Counts from good, overlapping photographs are much more dependable and a deal more comfortable.

The St. Kildans snared their birds with long, tapering poles from which dangled a noose made from horse-hair with a split gannet quill plaited in. Armed with these they scoured those terrifying cliffs clubbing thousands of lusty gugas amidst a frenzy of deafening calls from panic-stricken adults and high-pitched yipping from terrified chicks. It was sheer bedlam. And over all the stench of regurgitated fish, rotting seaweed and that subtle musty smell of seabird feathers. Gannets, fulmars and puffins provided meat, fat and feathers through the long, dark North Atlantic winters. Fulmars, which in the late nineteenth century nested only on St. Kilda were uniquely valuable to the islanders — "can the world exhibit a more valuable commodity?" (presumably they didn't know anything about camels.) "The fulmar furnishes oil for the lamp, down for the bed, the most salubrious food and the most efficacious ointment for healing wounds, besides a thousand other virtues —. But to say all in one word, deprive us of the fulmar and St. Kilda is no more." Slippers were made from the skin of the gannet's head and neck turned inside-out and sewn up at one end. In that way the soft neck feathers snuggled against their bare feet and the gannet's tough skin faced the outside world. They must have been supremely comfortable and their short life didn't matter; there were plenty more. Imagine the St. Kildans being the first throwaway society.

*Fulmars on St Kilda. It was from Kilda that fulmars began their spread around Britain*

They stored fish and seafowl in 'cleits'; small dry-stone structures with turf roofs which allowed the wind to pass through and prevented the contents from mouldering. These typical St. Kildan larders were, and still are, scattered thickly on Hirta's slopes outside the village. Cleits were not solely for food; they were used to dry crops and turves for fuel and on cool and windy old St. Kilda they worked well. Like the guileless North Ronans the St. Kildans, for nearly all the thousand or more years they occupied the island —"knew nothing of money or gold, having no occasion for either. They neither sell nor buy, but only barter. They covet no wealth, being fully content and satisfied with food and raiment". So wrote Martin Martin in 1703 in his "Description of the Western Islands of Scotland", referring to North Ronans.

The last of the amazing people of St Kilda finally and perhaps sadly left their ancestral home in 1930, almost within my lifetime. Their departure brought to an end more than a thousand years of continuous occupation of a harsh, demanding but dearly loved island home. A biographer of St.Kilda wrote—"they drowned the dogs in the sea, the cats were left alive" (a woefully bad idea on a seabird island) "in each cottage was left an open bible and a small pile of oats. Doors were locked". This small, hardy community had survived for more than a millennium with no welfare provisions except those they devised for themselves, no law-enforcers, no amenities or

health care and an unimaginably tough life. Yet, just like the North Ronans, they were a contented, crime-free, fully viable society. What are we to deduce from that? Small is beautiful?

It has recently been pointed out that the crofting practice on St. Kilda, combined with the islanders' diet of seabirds, caused certain chemical elements, notably lead, zinc, cadmium and arsenic, to accumulate in Hirta's soil but it is not known whether this affected their health and, because of this, their decision to evacuate the island. It seems unlikely. They were apparently pretty healthy for a thousand years.

*The village with cleats in which seabirds were stored as winter food*

259

*Exploring St. Kilda*

On May 26th 1989 a 32' yacht (a 'Rival') the pride and joy of Kirkcudbright's harbourmaster, Bill Morgan, sailed down the river Dee bound for St. Kilda, with a motley crew, none mottlier than me, though local solicitor Willie Henry ran me close. The wind was bang on our nose and obdurately remained so for every kilometre of the long slog up Scotland's west coast, past the Inner Hebrides and Jura's shapely paps and eventually to North Uist and out to St. Kilda, over 400 kilometres and 55 hours on the engine. We anchored in Village Bay late on Sunday night as the summer dusk softened the stark outlines of the islands, black against a lowering sky and leaden sea. "The islands had no depth or range of colour; they were razor-sharp monotone silhouettes like exaggerated card-board cut-outs stuck up on edge and lit from behind" (Robert Atkinson, 'Island Going' 1949, Collins). There were few signs of the millions of seabirds that nest there, gannets, shearwaters, petrels, puffins, kittiwakes, fulmars and skuas. From far out in the wilderness of the North Atlantic they stream back to the cliffs, boulder-slopes and burrows for the brief but bountiful northern summer, millions of eggs laid, tens of thousands of adults lost to predatory skuas, greater black-backed gulls and accidents; the endless cycle.

Next day, eager to soak up the atmosphere of this seabird mecca we set off for the top of Hirta and the view out to Boreray and the two stacs. Famous though they are it was the formidable bonxie or great skua that left the most indelible impression on my two companions. Menacing, hoarse of voice, heavy and prone to strike unwary heads with low-level attacks from behind, the predatory bonxie has an evil and well-earned reputation. Skuas strike with their webbed feet, furnished with sharp claws; never with their powerful bills. Most folk find them highly intimidating but despite tales of split scalps, in my fairly limited experience they are pretty harmless. Still, a hat or a stick hoisted above the head can be handy.

We quit Village Bay in some haste. A heavy Atlantic swell had begun to surge in so we got out quickly, weighing anchor at 5 a.m. and motoring the 70 kilometres across the Sound of Harris to Loch Maddy. The calendar said it was summer but up in the bows the wind pierced like a knife. My job was to keep a sharp lookout for substantial floating debris which could easily have holed us. What trawlermen must endure during freezing winter gales in the North Sea and North Atlantic. What agonies seamen suffered on the clipper voyages to London, from Australia via Cape Horn; furling board-stiff sails, frozen solid, 30 metres up in the rigging, half blinded by smoking gales of sleet and snow, with the clipper racing through huge seas like an express train. There was no rescue for anybody who fell overboard. To turn would have torn the sticks out of her.

Some years after this visit to St. Kilda I again anchored in Village Bay, this time in ex-Russian naval vessel 'Professor Molchanov'. On this benign early summer morning a large trawler attracted a missile storm of diving gannets to its discarded offal. In such a feeding frenzy they go mad, criss-crossing each other's path and cleaving the water within centimetres of those bobbing up after their dive. How do they miss each other in the melee? In fact they don't always; one gannet was found with a puffin necklace. Mike Harris, who has spent a lot of time on St. Kilda, saw a gannet shear off a wing against a trawler's rigging whilst the crew were gutting the catch under arc-lamps. Gurney, the original chronicler of the gannet records them diving suicidally into a fish-gutting shed in Cornwall. Gannets are wonderful but not wildly intelligent.

It was a rare treat to visit Foula in the Shetlands, now the most isolated inhabited island in Britain —" far on old ocean's utmost region cast, one lonely isle o'erlooks the boundless waste". Foula, some three kilometres by three, holds the site of a Celtic Broch and a bronze-age burial ground. It once held a large community but by the mid-sixties only some 37. According to Chris Mylne it was probably the last place in

*Trawler attended by gannets in Village Bay, St. Kilda*

261

Britain to use the old Norse version of the Lord's Prayer. Foula, or Fugly, means Bird Island. Its most awesome feature is the Kame of Foula, a cliff over 400m high. These days even remote old Foula is never far from help. Whilst I was anxiously watching an eider duck shepherding her brood of tinies across the landing strip, desperately trying to cope with the menacing attention of a predatory arctic skua the Inter-Island hospital plane came in to evacuate a woman patient. Even people who get into difficulties at the South Pole can be air-lifted out. Foula is the stronghold of the piratical bonxie. For decades Bob Furness and his students from Glasgow University's zoology department have assiduously followed the fortunes of Foula's skuas, discovering many ornithological nuggets, one of which concerns the huge number of non-breeding birds which come and go throughout the season. It is difficult to say why they are there and what they are doing. Difficult questions are common in ecology, which is why projects have to grind on for years. Even if you had time and resources you would still have problems with these skuas. Perhaps you could figure out a way of trapping large numbers and then fit them with tracking devices to follow them for a few years. You would never sell that expensive idea to a grant-giving body. Long-term studies of seabirds can be valuable barometers of change when things are going wrong for the birds but they are difficult, costly and time-consuming. Researchers tend to go for trendier, lab.-based projects with quicker pay-offs more relevant to human-centred interests. My sort of work would fare extremely ill for grants these days.

The bridge of 'Professor Molchanov' treated us to a close view of remote and austere Sule Stack (Stack Skerry or Stack of Suliskerry). Of all the North Atlantic gannetries it is in some ways the most remarkable. Guarded by swell from the open Atlantic and bereft of any half reasonable landing place this ancient gannetry is so inaccessible that only a tiny handful of naturalists have visited it in the last half-century or more. Even more unusual, the "full-up" sign has been hanging in the window for donkey's years. As long ago as 1887 gannets densely covered the entire surface. Thousands of youngsters from this forbidding stack of naked hornblende gneiss have been forced to seek nesting space elsewhere. That, indeed, is exactly how most gannetries grow, by sucking in recruits from other colonies. 'Sule Stack' figures high on my 'wanted' list but it is so out-of-the-way that I hadn't expected to visit it. Gannetries like the Bass and Ailsa are one thing; the Sule stacks of this world another. From the ship's bridge I delivered my finest eulogy to a remarkably disinterested audience; the communication system had been switched off.

'Sule' is old Icelandic for 'gannet' and the name is by no means unique to Sule Stack. It has been bestowed, too, on the far more famous gannetry of Sule Sgeir (over 70m high and a few score square metres in area) which we also visited. For centuries the main Scottish gannetries such as Ailsa Craig, Bass Rock, St Kilda and Sule Sgeir were plundered every year for their crop of fine, fat gugas, as well as for adults. Until outlawed late in the nineteenth century this annual massacre of many thousands kept the British population of gannets artificially low. Today only one gannetry legally suffers this slaughter: Sule Sgeir, 70km NE of the Butt of Lewis.

In 1938 Robert Atkinson wrote: "the passage in an open unengined sailing boat used to take anything from 10 hours to a couple of days. This boat, the 'Peaceful', was

*Sule Sgeir (top) and Sule Stack (above) famous remote gannetries*

a Ness Sco, a shallow draft double-ender of 18′ keel, in the lineage of Norse galleys. She was very nearly the last of her sort left in the Hebrides and had been built in Port of Ness in about 1913 by the Macleods, the only boat-builders in the Outer Isles …. When they got to the rock they hauled the boat up high and dry 60′ up the cliff. They cleared the shags and muck out of the old bothies and lived in them for up to three weeks. In that time they would expect to kill, pluck and salt down round about 2000 gugas, about 6 tons of them. The meat of the guga was black and had a thick layer of white fat over it. When they got back to Ness they would be posting the gugas all over the world, USA, New Zealand, Australia, Canada, pretty well wherever there were Lewis men. The gugas' circulation, like the 'Stornoway Gazette', was worldwide".

The Lewismen lived in bothies 'beehive shielings' made of overlapping drystone flags stepped inwards and covered with turf. At the time of Atkinson's visit the gannetry was decreasing southwards, away from the centre of the rock, and the areas which the gannets had vacated turned green with thrift and orache, dotted with breeding fulmars.

Atkinson makes the mistake, often repeated, of assuming that as a normal practice gannets regurgitate fish onto the ground for their young to pick up. He writes "the gannets kept on sicking up fish all day long". No doubt they did whilst he was there for they were desperately afraid. Vomiting is the normal reflex response to fear, an adaptive reaction which lightens them and makes it easier for them to take off. And if gannets had reason for fear anywhere, Sule Sgeir was the place. They had been clubbed to death for centuries. In nearly 50 years, I have never seen an undisturbed gannet deposit fish on the ground, although sometimes the chick lets it fall during transfer from bill to bill.

As I said, even today a few hardy Lewismen sail across to this forbidding Rock, pull their boat up clear of the surge, refurbish the old bothy and fall to their grim task. Arduous, dangerous and one would think sickening work clambering around the hazardous faces clubbing the panicking youngsters. Then comes the laborious and seemingly endless task of plucking and singeing 2000 gugas, wearisome toil. The number is supposed to be fixed at 2000 though seldom if ever checked. In addition, an uncounted number are displaced, probably injured, and die miserably at sea. Some of them are still partly clothed in down, but in any case, if forced to sea prematurely they are doomed. Nowadays this gory affair is unnecessary for food or profit, although understandably Lewismen feel strongly attached to the expedition, perhaps especially in this technical, over-sanitised and risk-averse age. So, for as long as Lewismen want to practise their ancient prerogative it will probably continue. Incidentally, the fact that gannets on Sule Sgeir have actually increased during the last fifty years or so comes from immigration and in no way implies that culling young gannets benefits the Sule Sgeir population. Let's get that straight before somebody makes a fatuous claim.

Before we left Sule Sgeir we rescued two gannets tied together with that devilishly tough synthetic fish netting. Over the years hundreds of the youngsters that I ringed on the Bass perished in this horrible way, a slow and cruel end, and yet another way in which man's pollution of the sea is paid for by other species.

On a beautiful May morning 'Professor Molchanov' dropped anchor off yet another 'Sule', this time Sule Skerry, a seldom visited seabird mecca and, incidentally, one of the most recently established gannetries, established in 2003 but since pipped by the Noup of Westray. Not inevitably will a small new 'colony', if a very few pairs be dignified by this term, survive and grow, but once it has passed a critical point it will almost certainly persist. At the time of our visit to Sule Skerry there were no gannets but the island pulsated with other seabirds, especially terns and puffins, the air electric with the exciting tumult of a seabird colony in full swing. As we zoomed happily along in the zodiac maybe 50m offshore, soaking up the exhilarating atmosphere, a large Atlantic roller caught the helmswoman napping. She was a fraction too late turning in to it so as to take it head-on, and it hit us broadside. The dinghy tipped onto its side and neatly catapulted an elderly American couple overboard. Two be-spectacled faces peered up like seals, from a deep Atlantic trough, whilst an agitated crew rushed to the side to heave them back on board. They were a dead weight ; I couldn't believe how heavy. It took two of us all our strength to drag each of them back on board. Everybody was soaked and a lot of expensive cameras ruined with salt water; I imagine that the resulting claims caused a few raised eyebrows in the insur-ance industry. It was a useful, and luckily a cheap lesson.

North Rona! Is there a more exciting Scottish island? "O' those endless little isles! And of all these little isles, this Ronay. Yet, much as hath been seen, not to see thee, lying clad with soft verdure and in thine awful solitude, afar off in the lap of the wild ocean, - not to have seen thee with the carnal eye will be to have seen nothing" . The chapel on Rona was "dedicated to St. Ronan, son of Berach, a Scot mentioned by Bede as having had disputations with Finan, Bishop of Lindisfarne, about the true date of Easter, and who, near to the close of his life, is said to have retired to the island of Rona and there died about the end of the seventh century" (T.S. Muir). Many years ago I revelled in Frank Darling's account of his life on uninhabited Rona with his es-timable wife, Bobbie, studying grey seals. I looked up to Darling as a person, a writer and a scientist and was dismayed and disillusioned when he forsook her. She sounds to have been a wonderful woman.

The last permanent inhabitant on Rona, evacuated in 1844, ended at least700 years of human occupation. St. Ronan's retreat was built in the eighth century but there is no certainty that even he was the first inhabitant. Like the St. Kildans the people of Rona, said to be simple and contented, had no interest in money or any other form of wealth. But tragedy came when rats got ashore from a visiting ship and ate much of their store of simple food. In 'The Scottish Islands' Haswell-Smith records that, almost beyond belief, the crew of a visiting ship came ashore and killed the islanders' only bull. Not long after this double tragedy the islanders left.

Despite its challenging exposure North Rona, at least on the fine summer day of our visit, seemed idyllic. Near the ruins of the ancient chapel several fulmars nesting at the base of the old walls, appeared totally careless of the predatory great-black-backed gulls breeding nearby. Few birds molest the oil-spitting fulmar, and with good reason. On the small island of Copinsay, a memorial to James Fisher, I came across a raven which could hardly fly. It laboured along on wings that seemed lead-heavy. It had been wrecked by fulmar oil, probably as the outcome of a squabble over the own-ership of a nesting ledge. Rona is now a National Nature Reserve, treasured for its thousands of Leach's fork-tailed petrels. These charming little tubenoses captivated the Darlings with their soft, ventriloquial churring. He said it would be a sad day when he could never again look forward to falling asleep with their mysterious noc-turnal song in his ears. In addition to its other treasures Rona is the stronghold of the grey seal; more than 7,000 haul out each year to drop their pups on the shores of this remote haven.

Fair Isle is a famous bird-watcher's Mecca and the site of perhaps the most renowned bird observatory in Europe, if that accolade be denied to Heligoland. It is the landfall for an astonishing variety of rare migrants to Britain, including wanderers from North America. The list of these weary little travellers reads like a 'Who's Who' of the bird world. The island lies roughly between Orkney and Shetland and 'Noble Caledonia' had placed it high on its list of desirables to be visited by its cruise ships. The surrounding seas are notoriously savage and before the advent of radar there were frequent shipwrecks on adjacent rocks. Despite this Fair Isle, like Shetland itself, has been occupied since at least 3,500 BC. It is a lively place with a constant flow of birdwatchers, more in a month than Foula receives in a year, largely because of the buzz engendered by its rarities.

'Noble Caledonia' takes its good relationship with the islanders seriously and on this particular trip we organised an impromptu beach barbecue for whoever wanted to come, both observatory staff and islanders. The local eider ducklings dived industriously amongst the straps of shiny brown kelp whilst we set up trestle tables, organised the charcoal for grilling steak and chops and prepared mountains of salad. This island-hopping cruise was a new experience for me and one which I probably would not have chosen as a passenger, but I give it full marks. It afforded people the opportunity to visit remote and difficult islands and isolated rocks which they could never have done on a conventional cruise in a large liner. We landed on a greater number of romantic bird meccas in one week than most people manage in a lifetime. St. Kilda is difficult enough but Rona, Foula, Fetlar, the Shiants, Copinsay and Rhum are magic isles too, to say nothing of Sule Sgeir, Sule Stac and Sule Skerry. Wonderful little Moussa, uninhabited since 1861, is the site of the finest broch in existence. These fortified stone dwellings built by the Picts perhaps 2,000 years ago as a defence against marauders are squat, circular towers with walls tapering from 4m thick at the base to about two metres thick at the top. Three metres below the top the walls slope

*Broch on Moussa, its walls are inhabited in summer by breeding storm-petrels*

outwards, making them virtually unclimbable. They are thick enough to accommodate a honeycomb of galleries, and just about impregnable. Within these thick walls steps climb up to a parapet at the top. The nicest touch of all, though, came from deep within the walls where, invisible and totally secure, storm-petrels were churring – surely one of the most evocative of all seabird sounds. What an astonishing change from the wilderness of the open Atlantic where these delicate little petrels spend the winter months —" pelagics, marathon fliers who, if they could, would no doubt even hatch upon the face of the waters, as indeed the old salts held was the habit of the stormies , the bringers of evil weather" (Atkinson). But the bit of his account I like best describes his technical wizardry when weighing storm-petrel chicks on Rona: the scales were constructed with a penknife out of packing-case wood, with nail bearings and two empty butter tins. Weights were coins and matches, afterwards converted by advanced arithmetic into grams, Atkinson's companion sitting with a tiny fluff of petrel on his knee, a thermometer tucked under its wing stump. Both of them crouched over the scales, one coaxing the chick to stay in its butter tin, the other manipulating the coinage, the petrel slowly oscillating against threepence plus 20 matches. The record increase in 24 hours was 12.8g; from fourpence plus 20 matches to five and a half-pence plus six matches.

Like almost the whole of the western seaboard of Scotland Moussa is a treasure and a delight. Scottish weather and the infamous midge, may its tribe endure, will surely keep it so.

*Great skua dive-bombing*

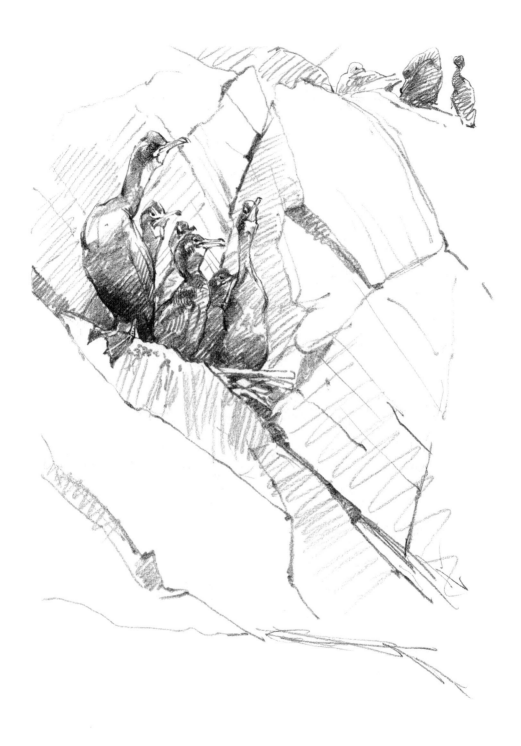

*Shag family on Craigleith Island*

Chapter 16

# THE CHANGING WORLD

*"Of course a Roman Emperor couldn't operate a comptometer or even a typewriter, but most modern executives can't either. He probably could operate an abacus, though, and it is doubtful if many executives could do that"*

Bergen B. Evans 1959 'The Natural History of Nonsense' Vintage Books New York

In the early sixties we had three hugely productive years on the Bass Rock – untold hours in a bitingly cold and draughty little canvas hide pegged precariously to the steep north-west face at the edge of the rapidly expanding colony. But see what science hath wrought! Now you can sit in the comfort of the Scottish Seabird Centre watching gannets in real time. I underline 'real time' because some visitors can't believe it. The picture is so good you can count lice on a gannet's head if they happen to be tempted out for a sip of fresh water by a shower of rain; little black specks that quickly dive back under the gannet down when the rain stops.

From the Bass we often gazed over the darkened Forth to the lights of the ancient Royal Burgh of North Berwick. We and the gannets seemed remote, cut off from the world. Not now. Today it seems nothing out of the ordinary to have live cameras on the Bass, transmitting pictures to screens in North Berwick, or for that matter elsewhere in the world. But twenty years ago the technology was not there and it took imagination even to envisage what we now take for granted. North Berwick's Bill Gardiner supplied the vision. The Scottish Seabird Centre was his brain child. Equally vital was the cooperation and enthusiasm of Sir Hew Dalrymple, whose family have owned the Bass since 1705. Now that the RSPB administer Grassholm and Ailsa Craig, amongst others, the Bass is our only major gannetry still privately owned, though the SSC now has a 99-year lease. Had the Bass not belonged to Sir Hew in 1960 I would have found it much harder to be allowed to live there. Dealing with bureaucrats is rarely as simple as agreeing with an individual.

Had I not been involved with the Scottish Seabird Centre from the outset I could never have imagined the difficulties besetting its birth. Raising more than three million pounds was just the start. There was the design of the building, the complex internal layout (much altered as we progressed), the technical problems to do with transmission of live pictures from the Bass, the morass of legal and administrative matters that snowballed and the multitude of other details that engulfed us. The endless planning

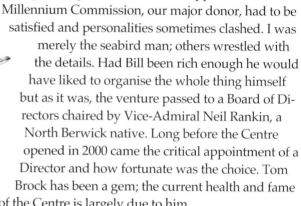

meetings multiplied, the deadlines approached, the Millennium Commission, our major donor, had to be satisfied and personalities sometimes clashed. I was merely the seabird man; others wrestled with the details. Had Bill been rich enough he would have liked to organise the whole thing himself but as it was, the venture passed to a Board of Directors chaired by Vice-Admiral Neil Rankin, a North Berwick native. Long before the Centre opened in 2000 came the critical appointment of a Director and how fortunate was the choice. Tom Brock has been a gem; the current health and fame of the Centre is largely due to him.

*Puffin display*

The success of the Centre could never have been guaranteed by mere men or even women. It was the handsome and charismatic gannet that did it. Gannets display spectacularly from February until October, long before and after the serpentine shags and gargling guillemots. As for old gaudy neb, Britain's favourite seabird, the puffin has a very mediocre repertoire. A bit of bowing and nebbing and that's about all. Puffins are so popular because they are clownlike and comical – toy birds. At the Seabird Centre you can delve into all their lives in a way that has never been possible before. You can focus the camera on whatever you want for as long as you like. Intriguing behaviour can be followed to a conclusion however long it takes. This is completely different from watching a nature film, edited by others. It is almost as good as being there.

*Construction of the Scottish Seabird Centre*

*The finished building*

The Scottish Seabird Centre is now a famous North Berwick landmark, attracting more than 50,000 visitors a year. At its heart are those live cameras, unblinking eyes that miss nothing. Some years ago, when many of the gannets were brooding small chicks, a terrific thunderstorm swept the Bass. Torrents gushed down the stony slopes now shorn of the soil and vegetation that used to soak up the water. It swirled around the sitting gannets nearly submerging the hapless chicks. The bewildered parents shook streams from their plumage and picked at the swirling debris. They are not programmed to deal with such an event, for there is nothing they could possibly do to help. The camera recorded the whole bizarre scene and the gannets' evident dilemma. Luckily almost all the chicks, tough little leather bags, recovered.

The unique advantage of the Bass camera pictures is that the gannet's behaviour is truly natural without any of the inhibiting human presence involved in conventional photography. In the Seabird Centre you are essentially sitting next to the bird. It is ideal for watching behaviour and some aspects of ecology such as breeding success and the time taken for chicks to grow. Remarkably, it would now be possible via satellite transmission to view live pictures from the Galapagos, events such as sibling-murder in the masked booby. Fifty years ago, on Ascension Island in the Atlantic, Doug Dorward, observed that the masked booby laid two eggs, but ended up with just a single chick, which had messily dispatched its sibling. A booby chick of three or four days is feeble, with a soft bill hardly ideal for murder, but like the new-born cuckoo which throws out its hosts' chicks, it has endless resolve. The parent pays no attention and the outcast is left to die. A camera on Tower Island could capture this drama and other marvels such as the Darwin's finch that uses a spine or a twig to winkle grubs out of cracks. And the Seabird Centre could have a satellite link with the

Falklands to watch penguins and albatrosses.

Since the Centre opened in 2000, many kinks have been ironed out, attractions added and others improved. One forgets the early traumas and the years of planning and execution. As usual, the devil was in the detail. The live cameras on the Bass needed a power supply but there was no electricity. The lighthouse lantern used paraffin so we opted for that to run the generator. But how to transfer a heavy generator and fifty-gallon drums of fuel from a small boat, tossing around on a sullen sea up to the lighthouse compound? In the old days there was a sturdy hand-cranked crane on the concrete landing platform and heavy items could easily be cranked up from a boat. Most of our supplies were landed in this way and in rough weather so were the keepers themselves. A steel cable, with a sturdy wooden box shackled to it, ran

*Lighthouse relief on the Bass, the helicopter was a late development after nearly 100 years of relief by boat*

steeply up from the landing to the lighthouse compound. Power came from a willing little donkey engine that "putt-putted" away imperturbably.

By the time the Seabird Centre became involved all this lifting gear had rusted solid or been dismantled. Since 1988 lighthouse reliefs had been by helicopter. Helicopter! That was the answer. The concrete landing pad was still in good shape though the path up to the lighthouse, running between the gloomy, fretted walls of the old Garrison buildings, was steep and narrow with awkward steps. But we could avoid all this by lowering the heavy items such as fifty-gallon drums directly into the compound, slung beneath the helicopter. It seemed simple!

As it happened, on the day chosen for this delicate operation the wind was fresh northerly. The helipad is on the south side of the Rock near the inner landing. Gannets invariably circle in hundreds above the sea on the side opposite to the wind direction. They then use the updraught created by the wind hitting the sea to give them height before they set off to fish or just to circle the Rock. So as we came in to land a dense column of circling gannets was right in our path. I cannot imagine that a helicopter blade could survive a collision with a 3kg gannet nor that a shorn helicopter could still helicopt. But our noncholant pilot evidently expected "those bloody birds" to get out of the way. I have never liked helicopters; unnatural things. Aeroplanes at least have wings.

*A helicopter is used to land equipment for the Scottish Seabird Centre*

With admirable aplomb and a devilish din as the noise ricocheted off the walls, a heavy generator and a fifty-gallon drum of fuel was lowered into the lighthouse compound. Alas, the generator in which we had put our faith gave endless trouble with the calamitous result that the cameras, our vital and much-trumpeted 'eyes' prying into the lives of the gannets were often out of action. This struck at the heart of the whole enterprise. Every time they failed it was embarrassing and difficult and expensive to get a technician out to fix it. Nowadays the cameras are powered by solar panels. Possibly we could have used wind power; it would have given me uncommon pleasure to see that accursed wind put to work.

Prince Charles officially opened the Scottish Seabird Centre in June 2000, Millenium year, and the weather gods clearly approved. The sun shone on enthusiastic crowds, bands played, bunting fluttered and the Prince ceremoniously unveiled the Foundation Stone by the harbour. Pat (now Macaulay) and Chris (Chris died in December 2012) represented the fourth generation of the Marrs, a notable North Berwick family who for many decades were the official boatmen for the Dalrymples as well as the Northern Lighthouse Board and they took a small party out in 'Sula II', the lovely mahogany open boat which operates in summer. We were attended by a motley flotilla including two fully crewed lifeboats. This may have been a genuine precaution in case we rammed a whale or perhaps simply an exercise for the lifeboats - doubtless a bit of both. The old Bass, never emotional, and often downright cantankerous, accepted the homage of yet another Royal – indeed another Charles, though this one not of the Stuart line as most of the other visiting Charleses had been. The worst of them was Charles II who incarcerated the poor old Covenanters. The gannets, the ultimate

*"Sula" with HRH Prince Charles, en route to the Bass Rock on the day in June 2000 when he officially opened the Scottish Seabird Centre. Neil Rankin, Chair of SSC Directors is on Prince Charles' right*

reason for all the fuss, paid no more heed than usual. If a strange boat visits the Rock they sometimes do swarm out and circle above it closely and in silence. They may behave in this strange way to any unusual object such as a floating tree-trunk, a wounded gannet or one entangled in fish netting; bizarre and puzzling behaviour.

So the Scottish Seabird Centre, officially born after a tricky gestation, now plays a huge part in the life of North Berwick and indeed in Scotland. It wins high praise and prestigious awards. Increasingly it is used in teaching and research by schools and universities. It should prove possible to monitor long-term changes in the timing and success of breeding and, importantly, the proportion of site-holding pairs that choose not to breed; interesting times for the gannet and indeed for other seabirds. The Scottish Seabird Centre is well-placed to play a big part in research and one cannot now imagine the old harbour without that soaring copper wing. What would St. Baldred have made of it? Or Hector Boece who wrote of the Bass and its "great store of Soland Geese"? But surely the gannets, more or less unchanged for a million years, will not be much different by the end of the next million. But will we, if we are still here?

*Guillemot and chick*

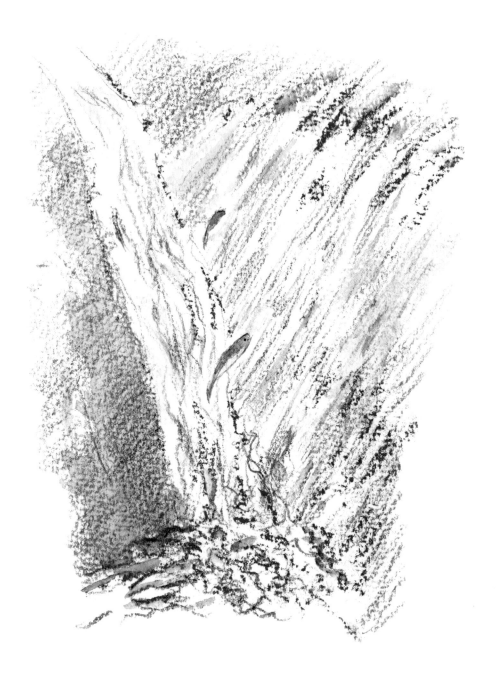

*Leaping salmon*

## Chapter 17

# BEGINNINGS

*"Collecting a haystack of material is one thing: orchestrating it into a narrative is another"*

John Wain 1991 Ed 'The Journals of James Boswell' Yale University

*"An island is a most enticing form of land"*

Russell King 1993 'The Geographical Fascination of Islands' Routledge London

My Britain of the 1930s seems light years away. no television (and not much radio), no computers, mobiles phones, refrigerators, pre-packaged food, supermarkets, milk in cartons or cheap flights to glitzy holiday resorts. How wonderful; but on the debit side some cars still had solid rubber tyres, batteries on the running board and dentistry was a nightmare. Milk came straight into the household jug from a milk can (1.5 gallons) with pint and 'gill' (half pint) measures hanging inside. As a boy, every weekend I delivered gallons of it, trudging up long garden paths, sneaking past belligerent dogs. It was hard work; 2 days labour for 30p. But in those days many a man with a family to support earned less than £10 a week. Carbide headlamps for cars were not long out of fashion and coal came by horse-and-cart, no miner at the coalface blacker than old Mr Swithenbank with his hundred weight sacks of best nuggets. He should have tipped them down an outside chute into our coal cellar but preferred to tramp through the kitchen. His hoarse, gritty call "co-ooal" sent my mother into a tizzy, laying down old sheets to foil his muddy boots. Each dusk, a lamplighter with a long pole did the round of the gas lamps in the streets. Very little had changed since the days of Queen Victoria whose head decorated many a coin still circulating. The old farthing, legal tender, was worth about an eighth of a present day penny.

Decades of industrial West Riding soot had blackened the golden sandstone of our three-storied terrace house. But for industry, it would have glowed like the Cotswolds. Its large, stone-flagged cellar kept milk cool in all but the sultriest summer weather. Our coal-fired kitchen range, religiously black-leaded, heated water to scalding-hot in an adjacent 'set-pot' for the Monday morning wash, 'mangled' between large wooden rollers powered by cast-iron cogs, a wooden handle and elbow grease.

Our kitchen sink, hewn from solid sandstone, made a fine bench for my father to 'cobble' shoes for us three boys. He cut new soles from a large sheet of shiny leather, but if the offending hole was small he just patched it. If the edges of the patch were not perfectly bevelled you tilted forward with each step but it kept our feet dry and with metal segs on the heels helped to eke out the money.

The long back garden opened onto a cobbled street lined with trees. Not your prissy planted specimens but mature trees, black poplar, ash, Lombardy poplar, horse-chestnut and sycamore. Oddly, there was not a single oak. It must have been a lovely area before it was built on in the late 1800s. In summer the trees cast a cool shade, and in Autumn drifts of fallen leaves hid the coveted shiny brown conkers.

There was plenty of open country nearby, good for birds. Ring ouzels, a fine indicator of unspoilt moorland, nested on Baildon Moor and in spring the cuckoos never paused for breath. I once counted more than 300 cuckoo calls in an unbroken stream. Meadow pipits parachuted to earth and skylarks poured out silvery cascades. Always foraging for nests, I found a pipits' with two cuckoo eggs. They are not difficult to identify for they are not exactly the same colour as their hosts' eggs and larger. I knew about the baby cuckoo's instincts and eagerly looked forward to the titanic battle between the two would-be murderers as they struggled to evict each other. I missed it, but I did watch the victor grow until it dwarfed the unfortunate foster parents. Its huge, insatiable orange gape goaded the harassed pair to super-pipit efforts, but at least they didn't have to stand on its shoulders to pop food in like one unfortunate wren foster-parent who over-balanced and was swallowed by its hungry baby.

*Wren feeding a cuckoo*

*Nightjar in flight*

These and a hundred other trivia left an indelible mark. Nightjars emerging on sultry July evenings, with their mysterious and ventriloquial 'churring' and staccato wing-clapping. How eagerly my brother and I waited as the dusk deepened over the valley, the sounds of human activity died away and then the magic moment when the air vibrated with that strange, unbirdlike sound. Or, arriving before dawn, we waited for the flitting form to pitch down in the vicinity of the nest, for in no other way could we find it. Tree creepers were another favourite, jerking their way up a trunk like clockwork mice to their nest hidden behind loose bark. Mallard rising in dawn panic from their guzzling in stream-side mud, the utter perfection of a lichened chaffinch nest on the bole of a silver birch. All these were something and nothing and yet everything. Smitten early, you are hooked for life.

The red West Yorkshire buses and the orange Penine service into the Dales opened up the countryside when it was too far to cycle. A few of the double-deckers had a winding stair to the upper deck totally outside the body of the bus and open to the weather. In the local playing fields, only half a km from our home, cricket filled our summer holidays. Failing that we could always prop up a dust-bin lid in the cobbled back street. A direct hit with a hard 'corky' ball had the impact of an exocet missile. I wouldn't swap such a boyhood for all the computer games in America. We were thin, fit and brown and our feet were hard. Looking back, it seems amazing that boyhood lasted so long. I wore short trousers well into my teens; today a seven-year old would die of embarrassment.

Everybody in Yorkshire – well maybe not recent immigrants – knows that Sir

Titus Salt built the industrial village of Saltaire to house textile workers from his gigantic mill on the banks of the river Aire. The Aire emerges as a crystal stream from the limestone near Malham Cove and ends, or it did in those days, as an open sewer, its banks and boulders lined with the sickly grey algae that thrives in sewage-polluted water. The school founded by Sir Titus saw me through the vital years between the County Minor exam at the age of 11 and the School Certificate at 16. His grimy lions, originally meant for Trafalgar Square but judged to be hardly grand enough for that exalted station, crouched regally on the pillars at our school entrance opposite a matching pair guarding the public library. Sir Titus was a philanthropist as well as a 'filthy capitalist'

During the wartime years of the early 1940s every inch of our school notebooks, with pages like woodchip wallpaper, had to be covered. The new-fangled ball-point pens were strongly discouraged because they destroyed the beautiful copperplate handwriting common to even the least scholastic of previous generations. A Jewish classmate with a rich bookie father, managed to get hold of a stock of biros which he sold surreptitiously. Zimmermann is probably a millionaire by now.

Our diminutive, walnut-wrinkled biology mistress used to nip out of the classroom for a quick gin-and-tonic, but she was an excellent teacher. Pistils and stamens, fruits and seeds, frog guts and earthworms were all grist to her mill. I loved biology. English fell to an attractive, raven-haired young woman, with a page-boy roll. She lived to 93 with a mind as sharp as ever. In her absence, the headmaster, 'Piggy' Parkin, took over, a burly, red-faced man with a broken nose and (doubtlessly related) an adenoidal voice. He had a penchant for teaching young boys the rudiments of boxing, which held no appeal for me. 'Speedy' (Mr Swift) the foghorn-voiced Chemistry master, was obsessed with the Periodic Table of Elements which at that time stood, I remember, at a modest 96, and with chemistry calculations which he worked through at full blast until the blackboard was covered with his neat, ruler-level handwriting and his lips caked with dried spittle. And there were memorable others; Mr Mathers (history), a Methodist local preacher kept pigs to help the war effort, and 'Josh' Gaskill (French) whose life revolved around his allotment.

School blazers and caps were braided with the colours of our four 'Houses'; Angles (yellow), Saxons (blue), Jutes (red) and Celts (green) which annually battled for supremacy on sport's day. It was far from a privileged school but had no problem with discipline or bullying.

World War II brought little hardship. There was indeed food rationing; an egg was a luxury and sweets and sugar gold dust, but it never worried us. We ate good home baking, bread, pies, teacakes, broth with sturdy suet dumplings that stuck to your ribs on a winter's day. Black-out seemed no problem and the occasional air raid a mere frisson of excitement – far different from the poor bomb-blasted souls of Hull, Coventry and London. When the eerie, undulating siren sounded we gathered with a few neighbours in our reinforced cellar until the long drawn out 'all clear' sounded, like a distraught cow. Bombs twice dropped within a couple of kilometres, demolishing part of Bradford, although nothing like as ruinously as modern developers have since done. Much of dignified old Bradford, built on the profits of the wool trade, has gone. So has the old market where I shopped for cheap dusty second hand books next to the pie and mushy peas booth.

So my boyhood was ordinary, happy and secure. We had caring parents, constancy, home cooking, a cosy fire in the bedroom when ill, a bike, a sledge, a tent for hot summer days in the garden, freedom from petty restraints and a wooden 'bogey' with pram wheels and brakes (a wooden bar pulled hard against the back wheels). But it was hard work for our parents. My mother washed all our clothes by hand, patched, darned, and ironed them with a heavy old flat iron heated on the kitchen fire. Our home was warm, bright and welcoming.

In my late teens the largely middle-aged members of the Bradford Natural History Society endured my tedious bird reports; 58 acorns recovered from the crop of a drake mallard, the order of the choristers in the dawn chorus, straw by straw accounts of the time it took for a pair of blackbirds to build a nest. Oh dear! It embarrasses me to remember the egocentricity of it all. It was pale stuff compared to the efforts of Messrs Briggs and Haxby in trapping, mounting and exhibiting hundreds of species of beautiful little moths (microlepidoptera), all meticulously displayed and labelled. I hope they eventually found a home in a Natural History Museum. Amongst those kindly naturalists was Sydney Jackson, curator of Natural History at Bradford's Museum – the Cartwright Hall. I passed blissful hours in his den at the top of the building and his benign influence landed me my first job as a fledgling analytical chemist in the laboratory of Bradford's pride and joy, the multi-million pound, state of the art, industrial sewage treatment complex at Esholt. My boss, the chief chemist and a keen birdwatcher, let me set up a drop-trap outside the laboratory. By pulling a string, I could operate it from one of the windows and ringed lots of common stuff - blackbirds, tits, robins and once a waterhen.

In the 19th and early 20th centuries Bradford was the thriving centre of the woollen industry. J.B. Priestley sketches the canny Bradfordians who travelled the world to buy wool in Australia and Africa, returning untainted by cosmopolitan sophistication, to gather in cliques outside the wool exchange, discursing in broad Yorkshire replete with 'thou', 'thine' and 'nobbut'. 'Thou' meant 'you' but only among intimates. A callow and presumptuous youth using it to an older man got his come uppance 'sither (see here) lad, tha thous them as thous thee first, think on'. What would these old Bradfordians think of their foreign city today?

Thousands of tons of greasy wool had to be washed before it could be combed in those vast, clattering satanic mills to which crowds of men, women and children scurried in the cold grey dawn to face a long day's tedious toil before returning to a grimy back-to-back with an outside toilet and no bath except a tin tub kept outside. Supper might be tripe, a sheep's head, trotters or fish and chips from one of the many excellent corner fish shops. Fish and chips for less than 4p in today's money was a real life saver for many. Meanwhile, the newly-rich wool magnates built fine mansions amongst rolling acres and tried to ape their social superiors. Now it is the turn of over-paid footballers, pop-stars and television personalities.

The wool washings poured into the local streams along with dyes and a lethal cocktail of other chemicals. The once crystal 'beck' which had rippled sweetly through the embryonic Bradford became 't'mucky beck' running red, blue or ochre. I crossed it every day on my way to primary school and it stank of wool suds and hydrogen sulphide.

Bradford City Council bought the delightful Esholt Estate from the Stansfield family, its rich woodland full of fragrant azaleas and rhododendrons. There was an avenue of holm oaks, a gracious old hall with a walled garden and fine greenhouses and hundreds of acres of farmland alongside the river Aire. Brilliant yellow wagtails nested in the lakeside meadows, kingfishers drew azure lines down the canal and rare lesser-spotted woodpeckers and tree pipits nested in the mature woodland. None survive now.

*Lesser spotted woodpecker*

The City's architects and engineers, of the ilk that built great railways, canals, bridges and portentous municipal buildings, took but a moment to bore a mighty underground sewer the several kilometres from Bradford to Esholt. They started from both ends and met flawlessly in the middle. Well, they may have been a centimetre out. To build a vast complex of tanks, filter beds and associated works (including a huge plant with lead chambers to manufacture sulphuric acid and a 70m mill chimney) capable of purifying millions of gallons of highly toxic industrial waste every day, took another eye-blink. Still Victorians, they did that sort of thing instead of messing about with computers and building wobbly bridges. And Britain still built great ships and dozens of different cars and motor-bikes; Norton, Matchless, Triumph, Scott, BSA and Royal Enfield. Visiting dignitaries came from as far afield as Japan. They were shown the foaming weir over which the purified sewage poured in a shining arc into the Aire and invited to drink a glassful. None ever accepted but a healthy pair of swans bred on the lake.

Esholt was my happy hunting ground, winter and summer, dawn and dusk. Paddy, an Irish night watchman rarely separated from his cloth cap and roll-up, sold me my first pair of binoculars. They came, god knows how, from the German navy, first world war, heavy old Zeiss with rubber eye-cups and individual eye-piece focussing. Paddy got them from a pawn shop and he sold them to me for £7. The first bird I really saw through them was a common sandpiper curtseying on a boulder in the clear, fast-flowing river Wharfe at Bolton Abbey. Those old glasses went with us to the Galapagos and Christmas Island and now live in fretful retirement, usurped by brash new Optoliths. It proved an early blessing having no binoculars for it made me identify birds by their calls and song.

During my teens birds took up all my spare time. Ailsa Craig, Holy Island, the Farnes, Mull, the Treshnish Isles, Cairngorms, Spurn Point and the local woods, moors and reservoirs figured large. Even in mid-winter I occasionally cycled a round trip of 90km on a heavy old Raleigh to see smew, goldeneye and goosander on a favoured reservoir. It was all birds, so it seemed natural that I should want to study zoology. And where could hold a candle to St Andrews, the oldest and finest university in Scotland? Its ancient grey stones are steeped in history; the ruined cathedral, venerable harbour, crumbling pier and frowning cliff-top castle with its bottle-dungeon. Houdini couldn't have escaped once thrown into that and the top sealed with a boulder. Golden sands stretched away to the Eden estuary and beyond to the wilderness of Tentsmuir Sands and Forest. During the crisp sunny autumn of 1955 I settled blissfully into 'Sallies', St Salvator's Hall of Residence. At dusk the skeins of grey geese flighted over to safe roosts on the Eden sandbanks.

The lab practicals were held in the Bute Medical Building. Afterwards we meandered back to Sallies past the centuries-old hawthorn propped up outside St Mary's Theological College, under the stone archway and across South Street and cobbled Market Street with its ancient Mercat Cross. The generous tea spread out on the long pine tables in Sallies dining room would have appalled a health faddist but it satisfied hungry students; bread, butter, jam, scones, pancakes, buns and tea; and then the proper dinner at 7 0. We were grossly pampered. During autumn and winter the housemaids laid a coal fire in every student's room, made the bed, dusted and cleared around. We simply lit the fire, drew the curtains and it was home from home; a far cry from the lot of today's students. University life, especially in ancient and prestigious ones like St Andrews, was a privilege (in my case totally paid for by a scholarship) confined to a small fraction of school leavers, and a good degree in a real subject more or less guaranteed a decent job. Even the 8 o'clock lectures on a dreary winter's morning felt part of an enviable life as we straggled across town, drawing our thick, long red gowns around us to thwart the east coast's snell winds. But Dr Macdonald's lectures on the reproduction of fungi seemed beyond tedium as he struggled apoplectically to make sense of mitosis and meiosis.

Science students worked hard. Lectures and long practicals ate up our time. Maybe we did feel a cut above the scribblers of the Arts and Humanities but they prosper well enough in later life, especially in law and politics. Douglas Cullen, a friend in Sallies, became Scotland's First Law Lord. Dear John MacGregor gave us privatised railways and left others to clear up the mess. Even in those days he was deeply into Conservative politics, often traipsing down to London to see Lord ('Bob') Boothby, of the hooded eyes and suede shoes; 'the poor man's Churchill'. Boothby was fond of a drink and when in his cups quite handy with crude mediaeval English. In 1959 he was voted in as St Andrews' Rector and I had the privilege of pushing him down the entrance steps of Sallies in a pram. Sadly the wheels stayed on.

Whilst I was still at St Andrews my oldest friend, John Busby, whose bird paintings adorn the walls of, among others, Prince Philip and Michael Heseltine, sold me his lovely old maroon and black 1936 Austin 12 for £20. It had a walnut fascia and its leather upholstery smelled deliciously of apples which were always kept in the capacious door pouches. Alas, one bitter winter's day after climbing the wickedly steep Devil's Elbow near Braemar (now by-passed) it literally shed its pistons in Newburgh's main street. An elderly Scot, surveying the ironmongery in the road with a keen eye, delivered his measured assessment "Nae laddie, ah doot ye've had it". And all because a wee pin came adrift and the crankshaft punched a hole through the cylinder block. It meant a complete replacement engine, by then a tall order, but my resourceful brother managed to find one in Yorkshire and the sturdy old Austin survived many more close shaves. The worst was in 1962 when it went on fire whilst garaged in a cowshed next door to Sir John Thompson's prize jersey herd at Stanton St John near Oxford. That winter was notably savage and to keep the radiator from freezing I had placed a small paraffin heater under the engine. It sounds ridiculous but it was accepted practice and I had used it countless times. However, on this occasion it must have flared up and set fire to the wiring. Providentially I had draped a heavy old Abercromby overcoat across the bonnet and it suffocated the fire. It cost £11 to have the car rewired in a back street Oxford garage. Many years later, after long mouldering in a shed, the old Austin was restored to mint condition by its new owner, won first prize in its vintage car class and was offered back to me by his widow for £10,000. I didn't have £10,000.

*1936 Austin 12 after a long lay-up*

*The 1936 Austin 12*

During my time at St Andrews I spent a couple of memorable summers in Deeside helping Pitlochry's Freshwater Biology Lab with its work on salmon. We trapped the mature fish as they swam up a tributary of the Dee to spawn. Willie Shearer, who directed the project, stripped out the eggs and milt and then planted thousands of fertilised eggs in the bed of a small stream, separated from the main river by a fish-proof barrier. When the parr were big enough to migrate downstream, they were obliged to pass through a collecting box and from the number so trapped Willie could calculate the percentage of eggs that had produced parr fit enough to migrate to the open sea. Simple information perhaps, but devilishly hard to obtain.

When working on salmon I camped in a derelict cottage at the Linn O'Dee and bathed in a roadside stone trough of icy spring water but my material well being depended on a marvellously hospitable highland family, the Grants. Each evening after a pleasant day 'up the water' Willie and I repaired to their welcoming table. Stag's liver fried in a blackened cast-iron pan, home cured ham, thick broth, fresh scones with home-churned butter and honey from their hives. The croft, set back from the road, was guarded by kamikaze bees and a tame but malevolent crow which terrorised unsuspecting visitors by landing on their shoulder and ferociously attacking an ear, with hilarious results for the onlooker. In the Grants' parlour, for its party trick, it tramped up and down on the piano producing, to my unsophisticated ears, something rather like Schoenberg. Mrs Grant's elderly father, a retired dominie, tiny and fresh faced at over 80, lived for the Sundays in summer when the royal family came on holiday to Balmoral. He was an elder at Crathes church and his boots fairly shone when they attended.

St Andrews gave me four memorable years, my zoology degree and (I was very proud of this) the D'Arcy Thompson medal. D'Arcy, the legendary polymath who wrote the classic 'On Growth and Form' was a former Professor of Zoology at St Andrews. He could equally well have filled the chair of Mathematics or, for that matter, Greek. These days it is hard enough for one mind to encompass a single major aspect of Zoology. But St Andrews failed to make me sensible. Bluff, ruddy Prof Callan, an authority on newt chromosomes, Jimmy Dodd an endocrinologist particularly partial to dogfish pituitaries, 'Daddy' Burt, a traditional Zoologist of encyclopaedic knowledge and Adrian Horridge, a brilliant neurologist filled me with whole hearted admiration wholly free from envy. I wanted none of it. I wanted to study animal behaviour like Tinbergen and Lorenz, and St Andrews was not interested. Prof Callan told me bluntly "Go away, this department is too small to do everything". St. Andrews didn't want me but Oxford and Cambridge were big enough to be kinder. There was friendly rivalry between Cambridge's Field Station at Madingley and Oxford's Zoology Department which included the Edward Grey Institute of Field Ornithology. Oxford leant more towards field studies and Cambridge, under the distinguished professors Thorpe and Hinde, leant towards experimental ethology using captive birds and mammals, so I went to the EGI.

As most Palestinians would agree, Lord Edward Grey of Fallodon made a better bird watcher than Foreign Secretary for Grey's deal with America, the Balfour Declaration during the 1914-18 war, eventually seeded Jewish immigration into British-controlled Palestine and look where that has led. The EGI under David Lack shared a building within the Botanic Gardens with Charles Elton's Bureau of Animal Populations, but the door connecting the two departments was kept firmly locked. I never understood why they seemed so territorial especially since some BAP staff were extremely birdy. In fact, Mick Southern at that time worked on tawny owls in Whytham Woods. But universities can be a bit odd and both David Lack and Charles Elton, though highly distinguished ecologists, seemed distinctly idiosyncratic.

The EGI houses the world famous W.B. Alexander Library, the finest collection of ornithological books and journals in Britain. Chris Perrins, Lack's field assistant, spent his days humping ladders around Whytham, checking the contents of innumerable tit boxes. His room, an appealing hotpotch of rings, equipment, half-eaten sandwiches, books and field gear couldn't have been more different from Lack's monastic cell. Reg Moreau, a wild fringe of hair flying around his bald pate and invariably sporting a green eye-shield, edited 'Ibis' the journal of the British Ornithological Union, perhaps an unfortunate title suggesting fratricidal strife against all who would put obstacles in the way of bird watchers. It is, in fact, a research-oriented bird journal. Bernard Stonehouse, Philip Ashmole and Doug Dorward, newly returned from the BOU Centenary Expedition to Ascension Island in the south Atlantic, claimed their few square metres of space. Father Lack regularly quizzed them for he had an insatiable thirst for information and a remarkable ability to concentrate his entire attention on the one topic that, at that time, interested him. Bill Bourne, the Bernard Levin of the academic bird world, studied bird migration using the new-fangled tool of radar. All these researchers, together with a plethora of visiting scientists gave the EGI a real buzz. The American Charles Sibley, famous in academic ornithology for his pioneering work on bird systematics using egg-white analysis, stayed for a year and all sorts of people dropped in for the ritual of afternoon tea. Father Lack presided whilst Christine, the charming secretary, poured. Never one for small talk, he steered the conversation along whatever line popped into his well stocked mind – drift migration, mediaeval church history, bread making or babies nappies all came up. Old fashioned and a trifle authoritarian, perhaps, but the EGI had a good family feeling.

I could not have written this book if I had accepted sound advice to do a nice, tidy, well-demarcated research topic for my final-year thesis at St Andrews. I could easily have joined two of my friends drawing stained slide preparations of lampbrush chromosomes of newts, for Prof Callan's obsessive study, or looking at bits of dogfish brain for Dr Dodd. Instead I chose to spend long hours coaxing reluctant male crested newts to launch into their spectacular courtship display and, when sufficiently turned-on, to deposit their spermatophore on the floor of their fish tank. Once that vital packet of sperm in its transparent gelatinous envelope had been laid the male had to entice the female to pick it up in her cloaca. She is just a poor little female newt who has no idea what a spermatophore is or does. So to tempt her to pick it up the male has to lure her to crawl over it. By 'pick it up' I don't mean using her hands, rolling over and stuffing it up her cloaca – nothing nearly so crude. If he can entice her to pass directly over it – no hesitation or deviation, but smack down the middle – the thing is simply sucked up by reflex; so quickly you can hardly see it. To this end, the male turns broadside on and presents the side of his tail which sports a snazzy white stripe, to the female. Then he sends wave after wave of undulations along its length whilst simultaneously moving away from the spermatophore. The hypnotised female, following that irresistible lure, passes over the spermatophore and hey presto, the job is done. This was an admirably perfect example of a complex, instinctive act and I was so lucky to have hit on it, especially since it had not been studied before. So, but for the newt there would have been no Bass Rock gannets, no Galapagos adventures, no Christmas Island jungle with Abbott's booby. Why did it all depend on the newt? Because the newt thesis persuaded Niko Tinbergen to take me on to study gannet behaviour.

Triturus cristatus awarded me a good enough degree at St Andrews to persuade Oxford's Edward Grey Institute to accept me for a D Phil and the erudite Dr Lack advised me to carry on with David Snow's elegant study on the ecology of the blackbird. So here I was, moodily chasing blackbirds, mere worm eaters, one step up from barnyard fowl, though they sing better. Every day I packed my mist nets, spring balances and notebook into the old Austin and set off soon after dawn from my lodgings in Woodstock Road to trap and mark blackbirds in gardens, fields and woods; all very well – quite good fun – but what could I add to David's meticulous study? No, Turdus merula was not for me, much though I loved the ambience at the EGI. So after one frustrating season in which magpies and crows ate all the eggs and chicks in my nests, I gazed enviously across at the Zoology Department where all sorts of exciting things went on. In a motley collection of tiny wooden hutches squatting on the flat roof of the main zoology building and accessible only by a ladder and trap-door, as well as out on the Farne Islands and amongst the golden dunes of Cumbria's Ravenglass, Niko Tinbergen's students rumbustiously engaged in studies of animal behaviour. The zest and informality stemmed directly from the maestro himself. Desmond Morris describes his first encounter with Niko. He had decided to leave his job as Curator of

*Black-headed gulls at Ravenglass*

Mammals at London Zoo to work on sticklebacks under Niko. Niko's room was frankly a tip, a hotch potch of fish tanks, photographic gear, piles of books and yel-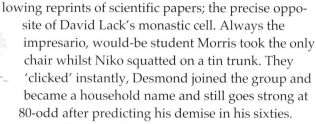lowing reprints of scientific papers; the precise oppo-site of David Lack's monastic cell. Always the impresario, would-be student Morris took the only chair whilst Niko squatted on a tin trunk. They 'clicked' instantly, Desmond joined the group and became a household name and still goes strong at 80-odd after predicting his demise in his sixties.

*Dipper*

When David Lack generously agreed that I could leave the worm-eaters to manage on their own and transfer to Niko, if he would have me, Niko, who knew absolutely nothing about me, had a look at my newt thesis (he'd studied newts himself) and decided I would make adequate gull fodder. So he offered me a place in his group of gull students. Ever since his early days amongst the sand dunes of the Dutch North Sea coast he had loved herring gulls. Sub-sequently his roving eye alighted on the black-headed gulls of the Ravenglass colony in Cumbria and every spring the Zoology Department's caravan migrated north to squat amongst the sand dunes and serve as the crumb-strewn, unwashed-tea-mug lit-tered common room for the tented encampment of gull watchers. Amongst them Hans Kruuk, Niko's Dutch student, eventually did magnificent work on hyenas and hunting dogs in the Serengeti and wrote a fine biography of Niko. Rather nervously I refused the offer to join the gull team and, instead came up with an idea of my own. I had vivid memories of experiences amongst the gannets of Ailsa Craig; the excite-ment, noise and hypnotic fascination of that tumultuous mass of seabirds. But it wasn't Ailsa I had in mind. My eyes were on the outer reaches of the Firth of Forth "the emerald-studded Forth" wherein lay the Craig "callet the Bass" in Hector Boece's words of 1527 "full of admiration and wonder, therein also is great store of Soland goose and nowhere else but in Ailsa and this rocke" (evidently he didn't know or re-member about St Kilda, which was already famous).

This Bass Rock whimsy of mine might have seemed wishy washy romanticism rather than a cool assessment of its research potential. Niko could easily have washed his hands of me and it would have been entirely my own fault. Most eminent academ-ics would have done precisely that. He hadn't chosen me from a long list of eager ap-plicants with carefully planned projects; he didn't work that way. I had just 'happened' by accident. I wasn't even an Oxford graduate and hopping from a presti-gious Institute like the EGI after less than a year was hardly a recommendation. He knew nothing about gannets and at that time neither of us had even set foot on the Bass to assess its suitability. There was no accommodation, unlike the situation on Ailsa Craig and nor did we know whether Sir Hew Hamilton-Dalrymple whose fam-ily have owned the Bass since the early 1700s would allow me to live there even if I could manage to build a hut on those steep, rough slopes. Looking back, it all seems so simple and trusting in a way one cannot imagine in today's rule-beset and safety-obsessed environment. There was no fuss, no bureaucracy, no legalities, no to-ing and fro-ing between Dalrymple and Oxford – just a 'gentleman's' agreement with no 'ifs', 'buts' or 'wherefores'. Nobody asked awkward questions about falling off cliffs. It was just a simple request on my part and an equally simple answer from Sir Hew and it worked perfectly; luck again.

There were, admittedly, a few snags. Where would I, soon to be 'we' live? What about food and water? How would the research be supervised? And how would prolonged isolation from the stimulus provided by other members of the Oxford behaviour group work out? There would be none of the interactions that had proved so useful to the Ravenglass team. Amazingly, in view of all this, I didn't even have to argue my case in detail with Niko, for which I have to thank Mike Cullen, Niko's second-in-command, who agreed to be my principal supervisor. And maybe, also, my newt thesis convinced Niko that I could work well on my own, for I had had no help during that project, and that I had a reasonable 'feel' for the subject.

*Mike Cullen*

Many of us who studied animal behaviour at Oxford during the 'fifties' and 'sixties', including some now famous such as John Krebs and Richard Dawkins (funny how we never called him 'Dick' whereas Richard Brown, also in the group, never got anything else; Dick Dawkins doesn't sound quite right) owe a great deal to Mike Cullen. Unconventional, sharp, modest, humorous, both highly cerebral yet practical to the point of caricature in avoiding frills and convention, and kind, especially with his valuable time, Mike was part of the fabric of our lives. My woolly romanticism drove him scatty. He had an old green Morris 1000 Traveller, the one with the wooden frame, and he treated it abominably. It retaliated by frequently breaking down, landing him in all sorts of awkward situations, especially on the Continent. Yet he seemed practical and good with his hands; quite capable of servicing it adequately. It seemed odd, and I often wondered if it could have been simply part of his philosophy. A car didn't matter much; just a few bits and pieces of metal – far simpler than a fish or bird – and it jolly well should work. He hadn't time to fuss with it. His own body, strong and hardy, came in for the same neglect and at least for lunch seemed to function largely on sliced white bread, margarine and raspberry jam, whilst spaghetti figured prominently for the evening meal. Unhappily, his heart eventually packed up, although one obviously can't say what part, if any, diet played. For Mike, eating seemed rather a waste of time unless combined with discussion of somebody's work or a behavioural topic. His total indifference to convention and protocol offended David Lack's finely-tuned sense of academic propriety, especially when Mike refused to pay £10 merely to have his D Phil confirmed; a dire affront in David's eyes. Mike hated ties and suits and however prestigious the occasion and his role in it, he considered his brick-red sweater good enough, although to cede a point he might grace an extra-special event by donning his undarned one. Perhaps unsurprisingly he was an agnostic or atheist, I'm not sure which, in contrast to the protocol-conscious, deeply religious David Lack. Nevertheless, of the two, Mike gave far, far more of himself and his time to his students and far less to his own advancement. Everybody thought the world of him

But back to the Bass. I still needed permission to live there. It looked pretty inhospitable, windswept and precipitous, without drinking water and uninhabited apart from the Lighthouse Keepers. However, St Baldred had lived there in the 7th century and so had a whole garrison of soldiers from around the 13th to the end of the 17th century. But where could we live? We abandoned the idea of clearing out the old dungeon below the garrison walls, now half full of detritus and commandeered by nesting fulmars. But there remained the roofless ruins of the 15th century chapel halfway up the south face on the presumed site of St Baldred's ancient cell. At one time it had been the garrison's ammunition store so it hardly seemed profane to put a modest cedarwood hut within its sheltering walls; a home for two impecunious gannet watchers. Ever helpful, Sir Hew agreed on condition that after the three years we would remove the hut and leave everything as before. I say 'we' because June hoped to come too, for we married on Dec 31st 1960. January may not be the ideal time for a honeymoon in Britain but the Lake District shone; brilliant in snow. Great Gable, thickly ice-crusted, gave us some excellent bum slides and we found the sort of guest house you can only dream about. Wood House, between Buttermere and Crummock Water was run by Miss Burns and the food beat Michelin 5-star into a cocked hat.

So, early in 1961 the ruined chapel on the Bass became our home for three memorably happy years. Better still, as this book has portrayed, it became the springboard for island adventures far afield in pursuit of tropical gannets, though at the time we had no inkling of that.

*Meadow pipit chasing a cuckoo*

# Forthcoming Titles in the Wildlife and People (W & P) Series. (formerly B & P)

1  Eagle Days - Stuart Rae (B & P)

2  Red Kites - Ian Carter

3  Short-eared Owl - Don Scott

4  On the Rocks ( sea birds) - Bryan Nelson and John Busby

5  Harriers and Honey Buzzards - Mike Henry.

6  Derek Ratcliffe - Various authors

Others to follow in the near future.